George Washington Plympton

How to Become an Engineer

George Washington Plympton

How to Become an Engineer

ISBN/EAN: 9783744646512

Printed in Europe, USA, Canada, Australia, Japan

Cover: Foto ©berggeist007 / pixelio.de

More available books at **www.hansebooks.com**

HE VAN NOSTRAND SCIENCE SERIES.

18mo, Boards.   Price 50 Cents Each.

Amply Illustrated when the Subject Demands.

No. 19.—STRENGTH OF BEAMS UNDER TRANSVERSE LOADS. By Prof. W. Allan, author of "Theory of Arches."

No. 20.—BRIDGE AND TUNNEL CENTRES. By John B. McMaster, C.E.

No. 21.—SAFETY VALVES. Second Edition. By Richard H. Buel, C.E.

No. 22.—HIGH MASONRY DAMS. By John B. McMaster, C.E.

No. 23.—THE FATIGUE OF METALS UNDER REPEATED STRAINS. With various Tables of Results and Experiments. From the German of Prof. Ludwig Spangenburgh, with a Preface by S. H. Shreve, A.M.

No. 24.—A PRACTICAL TREATISE ON THE TEETH OF WHEELS. By Prof. S. W. Robinson. Second edition, revised.

No. 25.—ON THE THEORY AND CALCULATION OF CON-TINUOUS BRIDGES. By Mansfield Merriman, Ph.D.

No. 26.—PRACTICAL TREATISE ON THE PROPERTIES OF CONTINUOUS BRIDGES. By Charles Bender, C.E.

No. 27.—ON BOILER INCRUSTATION AND CORROSION. By F. J. Rowan.

No. 28.—TRANSMISSION OF POWER BY WIRE ROPES. Second edition. By Albert W. Stahl, U S N.

No. 29.—STEAM INJECTORS. Translated from the French of M. Leon Pochet.

No. 30.—TERRESTRIAL MAGNETISM AND THE MAG-NETISM OF IRON VESSELS. By Prof. Fair-man Rogers.

No. 31.—THE SANITARY CONDITION OF DWELLING-HOUSES IN TOWN AND COUNTRY. By George E. Waring, jun.

No. 32.—CABLE-MAKING FOR SUSPENSION BRIDGES. By W. Hildebrand, C.E.

No. 33.—MECHANICS OF VENTILATION. By George W. Rafter, C.E.

No. 34 —FOUNDATIONS. By Prof. Jules Gaudard, C.E. Second edition. Translated from the French.

No. 35.—THE ANEROID BAROMETER: ITS CONSTRUC-TION AND USE. Compiled by George W. Plympton. Fourth edition.

No. 36.—MATTER AND MOTION. By J. Clerk Maxwell, M A.

No. 37.—GEOGRAPHICAL SURVEYING; ITS USES, METHODS, AND RESULTS. By Frank De Yeaux Carpenter, C.E.

No 38.—MAXIMUM STRESSES IN FRAMED BRIDGES. By Prof. William Cain, A.M., C.E.

HOW TO BECOME

AN

ENGINEER,

OR.

*THE THEORETICAL AND PRACTICAL TRAINING
NECESSARY IN FITTING FOR THE*

DUTIES OF THE CIVIL ENGINEER.

THE OPINIONS OF EMINENT AUTHORITIES,

AND

THE COURSES OF STUDY IN THE TECHNICAL SCHOOLS.

BY

GEO. W. PLYMPTON, Am. Soc. C.E.

NEW YORK:

D. VAN NOSTRAND COMPANY,

23 MURRAY AND 27 WARREN STREET.

1891.

PREFACE.

In answering the question suggested by the title of this little book, the writer has felt throughout the embarrassment arising from the consciousness that in the minds of the great numbers who almost daily ask the question there is almost as large a number of pursuits comprehended in the inquiry. The title engineer is assumed by men engaged in many varieties and many grades of human industry, from the ambitious plumber's apprentice, or the engine-driver of a tug-boat, to him who plans and directs the construction of the most extensive public works. The attempt has, however, been made to define the proper limits of the application of the term "engineering," and to advise the young man who desires to become an engineer in the generally accepted sense of the term how to direct his efforts in such way as to make profitable use of his time.

HOW TO BECOME AN ENGINEER.

CHAPTER I.

INTRODUCTION.

ENGINEERING is the science of employing the physical properties of matter to serve the purposes of mankind. It includes also the useful application of the different forms of Energy. Two branches of the science are recognized, the distinction being based upon the ends to be served. If applied to advance the interests of mankind in a state of peace, it is called *Civil Engineering* ; if to serve the purposes of war, it is *Military Engineering.*

Rankine says : " The term Civil Engineering is applied to a wide and somewhat indefinite range of subjects, but it may be defined as embracing those applications of mechanics and of the arts of construction generally which belong to lines of transport for goods and passengers, whether roads, railroads, canals, or navigable waters ; to works for the conveyance of water, whether for drainage or for water-supply ; to harbors and works for the protection of the coast. All these kinds of works are combinations of structures and machines ; they comprise *structures* in earthwork, as cuttings, embankments, and reservoirs ; in masonry, timber, and iron, as bridges, viaducts, aqueducts, locks, basins, piers, and breakwaters ; they comprise *machines*, such as cars and locomotives, lock-gates, sluices and valves, pumping steam-engines, and dredging-machines. Their principles therefore consist, to a great extent, of the general principles of construction and machinery combined and adapted to suit the circumstances of each kind of work."

" But *Civil Engineering* involves also the art of laying out lines of transport, and selecting the sites for works in the best manner possible with reference to the features of the country, so as to secure economy in execution and working."

The water-supply and drainage of cities must be contrived by an engineer, and the cost estimated in advance of the construction.

The relative merits of different systems of conveying goods or passengers must be determined by the engineer.

Now, because the profession of civil engineer involves such a variety of scientific labor, it happens that many engineers devote their time and energies to some of the various departments mentioned above. Thus the *Mechanical Engineer* devotes his time to machines, to their construction, use, and efficiency; also to the construction and operation of steam, gas, and air engines.

The *Mining Engineer* is prepared to direct the various operations connected with the digging of coal or metallic ores from below the surface of the earth, and converting them at once to convenient forms for use, or transporting them to market.

The *Electrical Engineer* confines his work to a narrower field. He contrives the means by which electricity is generated, and conducted to the places where it is to be applied in producing light, heat, or power. Too many in this branch of engineering are unfamiliar with the general principles which underlie all practice of civil engineering. Sir William Thomson, in a recent address to a body of electricians, reminded them that, to become electrical engineers, they should first make themselves engineers, and then become electricians.

The *Hydraulic Engineer* estimates the water-power of streams and determines the proper location for mills and factories. He designs systems of water-supply and drainage for cities and towns, and estimates their cost. He also plans and directs the improvement of navigable rivers and the construction of canals.

The *Sanitary Engineer* is engaged chiefly with plans for draining and ventilating buildings.

The thoroughly educated civil engi-

neer is he who has been well grounded
in the principles which underlie the
practice of the surveyor, the mechanical
engineer, the mining engineer, the hydraulic engineer, the electrical engineer,
and the sanitary engineer, and has acquired some familiarity with the practical work of each.

Military Engineering, as before explained, embraces the science and art
of war. The student who designs to
become a military engineer devotes
his time to acquiring much the same
branches of science as he who is to become a civil engineer, but it is always
with reference to applying his science to
the operations of war. His applied science embraces the construction of forti-
fications, either temporary or permanent,
the works of attack or defence of for-
tresses, the construction and laying of
military bridges; also the reconnaissances
and surveys for military purposes, including the operations of armies in the field,
and the construction of those transient

works by which troops are protected in line of battle.

Military Engineering also embraces gunnery, military pyrotechny, transportation of military stores, and the renewal of destroyed forts and bridges.

In this book we propose to deal with the training of the civil engineer only.

It is undeniable that many engineers have become eminent without the advantage of the systematic training which is generally considered necessary, and have risen to eminence from obscure positions where they had apparently become permanently settled after having reached mature years. Watt and Smeaton were instrument-makers, Telford was a mason, and George Stephenson was an engine-driver. All these men were possessed of a special aptitude for the work they accomplished, and they all, moreover, devoted much time to study after having passed the age which we now consider most favorable for the acquisition of book knowledge. They lived, moreover, at a time when the necessity for a special training for engineers in the schools was not thought of.

Professor Reynolds, in a lecture on " Engineering as a Profession," said : " Those who, in spite of their early paucity of education—and there have been many—have, notwithstanding, acquired a fair working knowledge of what we may call the theory of their subject, have done it by undergoing the greatest hardships. They have spent years of weary work in acquiring what would have been a comparatively short and pleasant task if systematically attempted with due means. I once heard an engineer, now of great eminence, when speaking of the comparative facilities now afforded, say: ' With a great sum obtained I this freedom ; but you, like St. Paul, are free-born.' I doubt not that there are others who have given up as hopeless the attempt to understand comparatively simple matters, because it was wrapped in forms into the mysteries of which they had not previously been initiated. One of the reasons why preliminary education

was held of such little account has been
the character of the only education to
be obtained. These things, however, are
now altered by the establishment of special schools and the extension of the old.
All the more useful branches of science
are now within the reach of the student
of engineering, and in the forms most
suitable for him. So that, as a step towards understanding the theory of machines, it is not now necessary for him
to begin with the theory of astronomy or
the doctrine of chances. The miserable
form in which the only mathematics to
be obtained was wrapped has compelled
engineers to work out methods for themselves, and now that the demand for such
knowledge has increased, we find that
the first mathematicians in the land
have, so to speak, patronized our system.
And, as is natural, the extension of the
practical use of mathematics infused new
life into their study. Learned as it now
may be with a special view to the application of practical mechanics, a knowledge of mathematics and science is much

more useful than it was. But this is by no means all. This would not be much were it not that the accumulated experience of engineering work has to a great extent been systematized and reduced to a form capable of mathematical treatment.

"The case is, however, entirely altered. Whereas formerly the good—at all events, the immediate good—to be reaped from the higher theoretical studies for those designed for the calling of engineering was doubtful, now there is every inducement for it. Nor has this fact been lost sight of by engineers. To their credit it must be said that they have come forward liberally to provide their successors with that education of which they have avowedly experienced the want. There are now some fifteen colleges and universities in the country where not only can a knowledge of all the useful sciences be obtained, but where the application of science and mathematics to the work of the engineer is made a special branch of study. The advisability of such a course of training is even now sometimes called in question. But I think this doubt can only apply, and is only meant to apply, to such a course of study as constituting the sole education of the student, as displacing the practical training. Looked at in this way, the doubt is just; for, as I have said, no amount of theoretical education can give that certainty and facility which only come from practice. As an adjunct to the practical training—as a preparation for it—there cannot, I think, be the least doubt. The field of engineering has become so vast that it is impossible for any one to acquire anything like a complete acquaintance with it by practical observation. The actual work of which one can gain experience in the course of a few years is but small, even under the most favorable circumstances. And the only way to make use of such experience as a general training is to supplement it by reading, and thus to use it for the purpose of illustrating the application of general laws and principles.

"Even allowing the aid of books, the range of work which may fall to the lot of an engineer is far too large to be mastered by one mind, unless reduced to a system and to general laws. Thus all the multifarious forms of structures—buildings, bridges, wheels, roofs, etc.—which, if each one is to be treated as a whole, must be numbered by thousands; if divided and considered in their component parts, are found to consist of seven or eight simple structures; and the laws which regulate the use of these may be treated separately. Or, again, endless as are the varieties of machines, when divided into their elementary parts these are not found to number more than 100. And so we might go on.

"It is, then, clear what an immense advantage is to be gained by attacking this mass of knowledge in a systematic manner, such as that in which it comes before a student in his course through a college. This is, in truth, the only manner in which anything like a complete mastery can be obtained. To attempt it by private study is to work at a great disadvantage.

"The exact course of preparation which is best for a student of engineering to pursue, although it should be varied according to circumstances, seems to be somewhat as follows: Assuming, as in other professions, the age at which he is supposed to commence his career to be about twenty-one or twenty-two: having pursued a general course of education at school until he is sixteen or seventeen, he should then commence his special course. In this he must learn something of science and something of art ; but he must also learn how the one can be brought to bear on the other. Mathematics and the natural sciences must form an essential part of his study, but he must not expect to make himself completely master of either. To do this would occupy more than the whole time at his disposal. He must select those branches of those subjects which most

directly relate to his future work, and
leave the rest as he would leave a luxury.
The making of this selection is very difficult ; the temptation is always to attempt too much, and this ends only in
confusion. It is but a comparatively
small portion of these wide subjects that
can be usefully brought to bear on engineering, and to these he must necessarily
restrict himself. The methods of applying these sciences to engineering problems constitute a large subject, and one
that it is necessary for him to study; and
besides this, he will have to devote some
of his time to acquiring sufficient knowledge of the things to be done by engineers, on which to study the application
of his science. And then there are yet
those manual operations which are essential to bring his knowledge to a practical issue, and in which a long course of
training is necessary to acquire the requisite skill, such as mechanical drawing
and the use of measuring and surveying
instruments, the want of facility in the
use of which would prevent for a long

time the student from making practical use of his knowledge.

"To acquire a useful knowledge in these various branches of study will require three, or at least two years. The student will then proceed with his practical training, which should include as great a range of work as possible. In this he will find the knowledge he has acquired of very great help; he will recognize much that he sees, and be able to judge of the most important things to which to direct his attention. After such preparation he will learn more in one year spent in the workshop or on the works than in three without it, so that by the time he has completed his training he will have as much practical knowledge as if he had spent his whole time in the workshops.

"Of course it would be little short of affectation to pretend that, surrounded as we are with mechanical results, one cannot learn to produce the results with which he is familiar, unless he is first able to deduce them from elementary principles. This would be equivalent to asserting that an English child could not speak English until he had mastered the rules of grammar. But to teach a language without the aid of grammar is not only a waste of labor, but a sure means of producing an imperfect result, and this is equivalent to teaching engineering without science. Such is the hold which the study of natural science has taken on all classes, and such are the facilities for those in the lower ranks to rise, that it seems to be quite certain that if those who have the best opportunity of qualifying themselves as engineers neglect to do so in the highest manner, they will find their places filled by those who, while rising from below, have made better use of their opportunities."

CHAPTER II.

SYSTEMATIC COURSE OF STUDY IN THE SCHOOLS AND COLLEGES OF THE UNITED STATES.

THE course of study required to obtain the degree of Civil Engineer differs in the higher technical schools chiefly in the amount of study required, not in the character of the branches pursued.

The Rensselaer Polytechnic Institute, at Troy, N. Y., is doubtless the leading engineering school of this country. Its course of study is given below. The tabulated statement is prefaced in the catalogue with the following summary of the specialties of engineering work:

"It should be stated, perhaps, that Civil Engineering is understood to include Mechanical or Dynamical Engineering, Road Engineering, Bridge Engineering, Hydraulic Engineering, Steam Engineering, Electrical Engineering, Mining Engineering, and Sanitary Engineering. By reference to the programme of the course of study, it will be seen that the wants of students of Mechanical and Electrical Engineering have been considered and well provided for, and that, with the supplemental course in Assaying, recently introduced, together with proposed special extensions of certain portions of the course, the wants of the future mining engineer will also be reasonably well supplied.

" The studies of the course are designed to secure to all the graduates a professional preparation, at once thorough and practical, for the following specialties of engineering practice:

"The location, construction, and superintendence of public works, as railways, canals, water-works, etc.; the design, construction, and management of mills, iron works, steel works, chemical works, and pneumatic works; the design and construction of roofs, arch-bridges, girderbridges, and suspension bridges; the survey and superintendence of mines; the design, construction, and use of wind-motors, hydraulic motors, air-engines, and the various kinds of steam-engines ; the design, construction, and use of machines in general, and the determination of their efficiency; the survey of rivers, lakes, and harbors, and the direction of their improvements ; the determination of latitude, longitude, time, and the meridian in geographical explorations, or for other purposes, together with the projection of maps ; the selection and test of materials used in construction, and the construction of the various kinds of geometrical and topographical drawings."

To enter the lowest class of the Institute the student must pass an examination in the following : Geography; English grammar, including spelling; Arithmetic, as treated in the higher text-books; Algebra, through equations of the second degree; Plane geometry, first five books of Wentworth's geometry, or its equivalent.

The full course is as follows :

COURSE IN CIVIL ENGINEERING.

FOUR YEARS.

DIVISION D.—FIRST YEAR.

MATHEMATICS. — Wells' University algebra; Wentworth's text-book of geometry; Wood's trigonometry, analytical, plane, and spherical.

DESCRIPTIVE GEOMETRY.—Warren's elementary plane problems—*plates;* Warren's elementary projections—*theory and plates.*

STEREOTOMY.—Warren's drafting instruments and operations—*theory and plates.*

PHYSICS.—Atkinson's Ganot's elementary physics through acoustics.

FRENCH LANGUAGE.—Fasquelle's French grammar.

ENGLISH LANGUAGE.—Hart's English composition and rhetoric.

GEODESY.—Gillespie's chain and compass surveying—*theory and practice;* farm surveying —*practice.*

TOPOGRAPHICAL DRAWING.—Elementary drawing; topographical plans.

FREE-HAND DRAWING.—Elementary practice.

DIVISION C.—SECOND YEAR.

MATHEMATICS.—Higher algebra; analytic geometry.

DESCRIPTIVE GEOMETRY. — General orthographic projections—*theory and plates.*

STEREOTOMY.—Bridge drawing; shades and shadows—*theory and plates;* linear perspective—*theory and plates.*

CHEMISTRY.—Inorganic chemistry.

PHYSICS.—Heat; optics.

NATURAL HISTORY.—Botany.

FRENCH LANGUAGE.—Syntax of grammar, with exercises and writing from dictation; translation of scientific works; epistolary correspondence and conversation.

ENGLISH LANGUAGE.—Composition; elements of criticism.

GEODESY.—Plane table surveying—*theory and practice;* adjustment and use of field instruments—*theory and practice;* trigonometrical and topographical surveying—*theory;* trigonometrical surveying and levelling—*practice;* mine surveying—*theory.*

TOPOGRAPHICAL DRAWING.—Map of farm survey; colored topography—*plates.*

FREE-HAND DRAWING.—Sketches of tools, of the components of machines, of bridges, and other structures.

DIVISION B.—THIRD YEAR.

MATHEMATICS.—Differential calculus; integral calculus.

ASTRONOMY.—Descriptive astronomy.

RATIONAL MECHANICS.—Mechanics of solids; mechanics of fluids; mechanical problems.

STEREOTOMY.—Machine construction and drawing—*theory and plates.*

PHYSICS.—Electricity and magnetism—*theory and practice.*

NATURAL HISTORY.—Mineralogy and petrography; descriptive geology; technical geology.

CHEMISTRY.—Qualitative analysis; blow-pipe analysis; determinative mineralogy; practical chemistry.

GEODESY.—Hydrographical, topographical, and town surveying—*practice.*

TOPOGRAPHICAL DRAWING.—Contour map; map of hydrographical survey.

DIVISION A.—FOURTH YEAR.

ASTRONOMY.—Spherical and practical astronomy.

PHYSICS.—Thermodynamics; electrodynamics.

PHYSICAL MECHANICS.—Mechanics of solids—*friction,—strength of materials;* mechanics of fluids—*practical hydraulics,—practical pneumatics.*

MACHINES. — General theory of machines; description of machines; theory of prime movers—*steam-engines,—air-engines,—electro-magnetic engines,—hydraulic motors,—wind motors;* construction and location of machines; designs for, and reviews of special machines; measurement and estimate of power; weir, and other measurements of the flow of water.

CONSTRUCTIONS.—Equilibrium and stability of structures — *revetement walls,* — *reservoirs,* — *roofs,* — *arches,* — *girder bridges,* — *suspension bridges;* designs for, and reviews of special structures.

STEREOTOMY. — Stone cutting — *theory and plates.*

GEODESY.—Higher geodesy; projection of maps—*theory;* line surveying—*road surveys,* —*staking out for constructions.*

ROAD ENGINEERING.—Common roads; railroads; canals; tunnels.

THE STEAM-ENGINE.—Lectures; indicating and estimating the power of steam-engines; duty tests of water-works pumping machinery; compound and multiple-expansion engines.

METALLURGY.—General metallurgy; iron metallurgy.

TOPOGRAPHICAL DRAWING.—Plans, profiles, and sections of railroad surveys.

LAW.—Law of contracts.

This is the most complete engineering course afforded in the United States.

The course in the Pardee Scientific Department of Lafayette College, at Easton. Pa., is as follows:

CIVIL ENGINEERING COURSE.

FRESHMAN YEAR.

First Term.—Algebra (completed), Elements of Industrial Drawing, English, March's Method, French, Chemistry, Lectures on Health.

Second Term.—Geometry (completed), Surveying, Plane Problems, French, German, Problems in Division of Land.

Third Term.—Surveying, Field Work, Elementary Projections, Trigonometry and Mensuration, French, German, Analytical Chemistry.

Throughout the year: Declamations, Themes, and the Bible.

SOPHOMORE YEAR.

First Term.—Analytical Geometry (begun), Surveying, Field Work, Elementary Projections, Mineralogy, French, German, Study of Words, Trench.

Second Term.—Analytical Geometry (completed), Topographical Drawing, Botany, Zoology, French, German, Mineralogy.

Third Term.—Differential and Integral Calculus, Descriptive Geometry, Botany, Zoology, French, German, Determinative Mineralogy.

Throughout the year: Declamations, Themes, and the Bible.

JUNIOR YEAR.

First Term.—Descriptive Geometry (General Orthographic Projections), Triangular Surveying) Field Work, Adjustment of Instruments, French, Mechanics, Lithology, Practice with the Blow-pipe.

Second Term.—Physics (begun), Calculus (continued), Shades and Shadows, Road Engineering—Theory (begun), Colored Topography, Hydrographical Surveying.

Third Term.—Linear Perspective, Physics (completed), Analytical and Applied Mechanics, Topographical Surveying, Map of Topographical Survey, Road Engineering—Theory (completed).

Throughout the year : Declamations, Themes, written Debates, and the Bible.

SENIOR YEAR.

First Term.—Water Supply, Road Engineering—Practice, Plans, Profiles, and Sections of Road Surveys, Astronomy, Machine Drawing, General Theory of Machines, Anatomy and Physiology.

Second Term.—Stone Cutting, Machinery and Motors, Strength of Materials, Stability of Structures, Supply and Distribution of Water, Astronomy, Geology, Mineralogy, Political Economy.

Third Term.—Bridge Drawing, Foundations, Retaining Walls, River and Canal Improvements, Designs for and Reviews of Engineering Works, Bridge and Roof Construction,

Graphical Statics, History, Geology, Graduation Theses.

Throughout the year: Themes, Speaking, and Biblical Studies.

Graduates from this course also receive the degree of Civil Engineer.

A course in Mining Engineering is also provided, differing from the above in the practical work during the Junior and Senior years.

In the Massachusetts Institute of Technology the course of Civil Engineering is specified as follows:

FIRST YEAR.

First Term. — Algebra, General Chemistry, Chemical Laboratory, Rhetoric, English Composition, French, Mechanical and Free-hand Drawing, Military Drill.

Second Term.—Solid Geometry, Plane Trigonometry, General Chemistry, Chemical Laboratory, Modern History, English Literature, French, Mechanical Drawing, Military Drill.

SECOND YEAR.

First Term.—Surveying: Compass and Transit, Plotting from Notes, Analytic Geometry, Advanced Geometrical Drawing, Physics, Modern History, German, Spherical Trigonometry.

Second Term.—Levelling. Profiles, Elements of Topography, Differential Calculus, Physics, Physical Geography, Modern History, German.

THIRD YEAR.

First Term.—Road Engineering, Advanced Field Work, Topographical Drawing, Integral Calculus, General Statics, Physics: Lectures and Laboratory, Structural Geology, Constitutional History, German.

Second Term.—Railroad Engineering, Topography and Map Work, Kinematics and Dynamics, Strength of Materials, Physics: Laboratory Work, Historical Geology, Political Economy, German.

FOURTH YEAR.

First Term.—Bridges and Roofs, Railroad Management, Hydraulic Engineering, Sanitary Engineering, Strength of Materials, Topography and Geodesy.

Second Term.—Bridges and Roofs, Hydraulic Engineering, Sanitary Engineering. Specifications and Contracts, Applied Mechanics, Thesis Work.

The following is the outline of the course of Civil Engineering at the School of Mines, Columbia College :

First Year.

Trigonometry and Mensuration, Physics, Botany, Geometrical Conic Sections, Analytical Geometry, Descriptive Geometry, Chemistry, and an extensive course in Surveying.

Second Year.

Analytical Geometry, Differential Calculus, Graphics, Stereotomy, Study of Roads and Pavements, Sanitary Engineering, Practical Mining, Zoology, Applied Chemistry, Blowpipe Analysis, Integral Calculus and Mineralogy, Surveying during summer vacation.

Third Year.

Mechanics of Solids, Physics, Practical Astronomy, Geodesy, Properties and Use of the Materials of Engineering, Metallurgy, Geology, Strength of Materials, Drawing, Practical Geodesy.

Fourth Year.

Water Supply Engineering, Sewerage, River and Harbor Improvement, Hydraulic Engineering, Machinery and Millwork, Graphic Statics, Railway Engineering, Railway Surveying and Practical Geodesy, Engineering Designs and Drawing.

Other institutions in this country affording similar courses of study are Cornell University, Princeton College, Rutgers College, Lehigh University, Stevens Institute, Polytechnic Institute of Brooklyn, and Michigan University. A course in the night-school of Cooper Union affords the same mathematical training as in most of the above courses, together with physics, chemistry, geology, and astronomy, but without the practical field-work. It has proved sufficient in several notable instances to serve as a groundwork to a successful course in engineering.

CHAPTER III.

ENGINEERING EDUCATION IN FOREIGN COUNTRIES.

THE following abstracts are taken from a report to the council of the Institution of Civil Engineers. The report was completed from documents obtained from the various countries mentioned.

THE STATUS AND EDUCATION OF ENGINEERS IN THE UNITED KINGDOM.

In England the profession of engineering is entirely unconnected with the government, there being no state corps of engineers other than those attached to the army. It is open to any one to enter the profession, and to obtain in it any standing his merits may entitle him to ; and all the civil works of the country, whether public or private, are (with some few exceptions, where Royal Engineers have been employed) executed by private practitioners.

There is, further, in England no public provision for engineering education. Every candidate for the profession must get his technical, like his general education, as best he can ; and this necessity has led to conditions of education peculiarly and essentially practical, such being the most direct and expeditious mode of getting into the way of practical employment.

The education of an engineer is, in fact, effected by a process analogous to that followed generally in trades, namely, by a simple course of apprenticeship, usually with a premium, to a practising engineer ; during which the pupil is supposed, by taking part in the ordinary business routine, to become gradually familiar with the practical duties of the profession, so as at last to acquire competency to perform them alone, or, at least, after some further practical experience in a subordinate capacity.

It is not the custom in England to consider *theoretical* knowledge as absolutely essential. It is true that most considerate masters recommend that such knowledge should be acquired, and prefer such pupils as have in some degree attained it, and it is also true that intelligent and earnest-minded pupils often spontaneously devote themselves, both before and during their pupilage, to theoretical studies; but these cases, though happily much more frequent now than formerly, really amount only to voluntary departures from the general rule.

The theoretical knowledge which may, in these cases, be desired, is obtained either by private reading or by attendance at the scientific classes established at various educational institutions, some of which have made special provision for studies of this kind, as may be seen in the particulars given farther on.

The *practical* education in England is perhaps the most perfect possible, if the opportunities obtained during the pupilage are ample, and the pupil properly

avails himself of them ; for nothing can
give a student so thorough and useful a
knowledge of practical works as being actually engaged for a length of time upon
them in a really working capacity ; in
addition to which, the habits of business
and the familiarity with all subsidiary
arrangements, acquired in this way, have
a beneficial influence on the student's future career. This thorough proficiency
in practical matters tends largely to com-
pensate for—in many cases to outweigh
—the deficiency in theoretical attainments, and it is undoubtedly this, influenced in some degree by the natural
self-reliance and practical common sense
inherent in the English character, which
has given such a high standing to the
profession in this country.

THE STATUS AND EDUCATION OF ENGI-
NEERS IN OTHER EUROPEAN COUN-
TRIES.

In most parts of the Continent the status of civil engineers differs materially
from that obtaining in England.   In al-

most every country of Europe there exists a state corps of engineers, educated and supported by the government, whose business it is to construct and superintend the public works of the nation. Private practitioners are therefore excluded from these works, and have to find employment, as best they can, in private industrial enterprises.

France affords the most perfect example of this system. The Government Corps of Engineers exists under two divisions, viz., the *Ingénieurs des Mines* and the *Ingénieurs des Ponts et Chaussées*. The former have the highest rank, and are employed chiefly, but not exclusively, on mining operations, and works allied thereto ; the latter take the more general public constructive works, as their name implies.

There are several classes in each division, inspectors general, engineers in chief, and ordinary engineers, and promotion is partly by merit and partly by seniority. The total number at present is given at 783. They hold a good position in the country, have generally considerable ability, and include in their ranks some men of great eminence in mechanical science.

Members of either corps are allowed, on special application, to undertake private work on the railways of the country (all other private work being forbidden), and receive a sort of furlough for the purpose. But if their absence from their official duties exceeds five years they forfeit their position, and lose all rights appertaining thereto.

The cases, however, of government engineers taking charge of private works are not numerous, as there exists in France a large body of civil engineers independent of the government, and having no official status, who devote themselves to the requirements of private individual enterprises. These have to make their own way in the profession, and occupy, in fact, the same kind of position as engineers in England, except that they have a sort of official guarantee as to their education, as will be hereafter explained.

The education of foreign engineers is strongly contrasted with that in England in every particular. Practical training by apprenticeship is unknown; the education begins at the other end, namely, by the compulsory acquirement of a high degree of theoretical knowledge, under the direction, and generally at the expense, of the government of the country. Partly with this, and partly afterwards, there is communicated a certain amount of information on practical matters; but this is imparted in a way differing much from the English plan, and probably with less efficient results.

Thus, while the English engineer is launched in his profession with the qualification of a considerable practical experience, but with perhaps little or no theoretical knowledge, the foreign one begins with a thorough foundation of principles, but with a limited course of practice; a deficiency, however, which tends to correct itself with time.

The education of both the government and the private engineers is on the same system, though carried on in different establishments.

The government engineers must have been at first pupils at a large general scientific educational establishment called the École Polytechnique. Admission to this is by public competition, and the standard is very high—so high, in fact, as to exclude all but persons already well advanced. The education in this school is exclusively scientific and theoretical, and from it students are taken to supply not only the corps of Government Civil Engineers, but also all the scientific departments of the army and navy.

After a two years' course in the École Polytechnique, such young men as are candidates for government employment as engineers are drafted off, also by strict examinations, into two special schools for the two departments respectively, namely, the École des Mines and the École des Ponts et Chaussées, in each of which the studies last three years.

During the five years thus spent, the *theoretical* education given to the engineer is very complete, every branch of science bearing on his profession being taught him, and his proficiency being tested by the strictest examination at the end of the term. On passing the final examination, the pupil enters the corps he is destined for, and begins at once his official duty in the lowest grade.

The *practical* education of the pupil, though not so complete or so effective as in England, is by no means neglected. During the three years' study in the special schools, much instruction is communicated having a practical bearing; lectures, descriptions, and exercises being given very fully on practical matters, with the object of making the pupil familiar with the general nature of the works he will hereafter have to do with, and so preparing him for his future experience on them. To aid this, the pupils are sent, for a considerable portion of the three years, on "missions" to various public works in practical execution under the department they are to be attached to ; but whether during these missions they actually take part in the works going on, or merely make observations, and write accounts of what they have seen, is not clear.

At any rate, at the end of the term (being then twenty-three or twenty-four years of age) they are assumed to be capable of doing useful practical work in the lower grade, or third class, and are at once given employment, with pay, as supernumeraries, draughtsmen, or sub-engineers, on some important engineering work in progress, where they gain the further practical experience necessary to fit them for taking more independent positions.

The education, both in the Ecole Polytechnique and the subsequent special school, is mainly at the cost of the government, the pupil only paying small fees. This fact, and the provision for life which the employment affords, produce a very keen competition for the privileges, which keeps up a high standard of qualification.

The education of civil engineers practising privately is given in an establishment called the École Centrale des Arts et Manufactures. This was originally founded as a private establishment; but it was afterwards taken to by the state, and is now entirely under government direction. The instruction, however, is not professedly gratuitous, as the pupils pay moderate fees.

Admission to this school is open to all who can pass a strict entrance examination; but the applications always much exceed the numbers that can be received (about 200 annually), and selection is made of the best.

The course of studies lasts three years, and is generally of the same nature as that given to the government engineers, with the exception that mathematical attainments are not pushed quite so high. In the first year the instruction is theoretical only; in the second and third years theoretical and practical instruction are combined. Thus the school aims at representing a combination, on a less extended scale, of the Polytechnic and special schools of the government corps. It is, moreover, more general and practical in its nature, so as to prepare the pupil as much as possible for any of the varieties of engineering work that may fall in his way, or, indeed, for other occupations of a scientific nature.

After this course is passed through, a diploma of "Ingénieur des Arts et Manufactures" is given to those students who have passed the highest public examination, and a lower certificate of capacity to those who have simply satisfied the important points. These documents give no claim to any employment, but are considered such good guarantees of ability, that their holders seldom fail to procure paid employment soon after leaving the school. They begin, like the government engineers, in subordinate situations, and gain experience and position as they go on.

There is nothing to prevent any engineer from practising in France who has not been through any of the acknowledged schools, and self-made men of superior practical ability have often succeeded well ; but these cases form the exceptions to the general rule.

In Prussia, a corps called Master Constructors (*Baumeister*) are employed by the state, and are educated as follows :

Each officer must first have received a complete general scientific training in one of the ordinary schools or gymnasia of the country.

He must then be practically engaged for one year with one of the constructive officers of the state.

He is then admitted into a special government educational establishment in Berlin, called the Royal School of Construction (Königliche Bau Akademie), where he remains two years, the studies comprising all branches of scientific knowledge appertaining to engineering and architecture, particular care being bestowed on construction and drawing. He then passes the first state examination, and enters upon practical paid employment in a subordinate capacity.

After three years of this, he devotes two more years to study, and then passes a second state examination, when he is considered fully qualified for a government appointment in the higher grade, which he will receive as vacancies occur.

Thus the complete education of the government official engineer occupies in all eight years from the time of his leaving the preliminary school, of which four years are devoted to actual practice—a feature that appears to be general in Germany, and that remarkably distinguishes the German curriculum from the French one, and brings it more into analogy with the English; with, however, the very important addition of the theoretical acquirements. It has, in fact, the advantages of the English and the French systems combined.

It is peculiar to the Prussian government system that the student must fully qualify both in engineering and architecture.

There is also in Berlin a government school for private practitioners, called the Royal Industrial Academy (Königliche Gewerbe Akademie), analogous to the École Centrale of Paris. This has also the feature of requiring the education to be commenced by passing some time in practical employment. The course in the school occupies three years, and the certificate given is generally a sufficient recommendation to remunerative employment.

In the Duchy of Baden the arrangements for engineering education in the Polytechnic Institution at Carlsruhe are noted for their perfection, and in consequence the school is much frequented by foreigners. Copious information will be found as to this school; and, in order to convey a more complete idea of the nature of the education, there is added a complete list of the questions given for the examination for the Diploma in Civil Engineering in 1867–8. The degree of proficiency, both in theory and practice, required for the proper solution of these questions must be very remarkable.

The system in Austria seems pretty nearly the same as in Prussia, except that there would appear to be only one educational establishment, the Polytechnic Institute, for all classes of engineers, and that any students are eligible for government employment, on passing the required examinations. After examination diplomas are granted, guaranteeing the theoretical and practical proficiency of the student; and licenses to practise in engineering, architecture, and surveying must be obtained from the government, according to prescribed rules. This restriction on private practice appears peculiar to Austria and some neighboring states. In Prussia, and in most other German countries, as in France and England, the right to practise is free.

The system in Russia appears pretty nearly the same as in France.

In Switzerland, the Polytechnic School of Zurich and the Special School of Lausame bear a high character for engineering education.

In Italy there are also good educational arrangements.

In Spain there is a corps of government engineers somewhat analogous to those of France, and their education is properly provided for.

THE COURSES OF STUDY IN SOME OF THE LEADING EDUCATIONAL INSTITUTIONS IN GREAT BRITAIN AND IRELAND WHERE INSTRUCTION IS GIVEN BEARING ON THE PROFESSION OF ENGINEERING.

KING'S COLLEGE, LONDON.

This is one of the most frequented institutions for the preparatory training of young men about to enter the profession of Civil Engineering in England.

It comprises several departments of education, but the one which has more particularly to do with this subject is the "Department of Applied Sciences," which is thus defined by the authorities of the college :

"The object in view in this department is to provide a system of general education, practical in its nature, for a large class of young men who in after life are likely to be engaged in commercial and agricultural pursuits, or in professional employments, such as Civil and Military Engineering, Surveying, Architecture, and the higher branches of manufacturing art."

It is also intended to prepare students for scientific examinations, such as those of the University of London, the Department of Public Works, India, Whitworth Scholarships, etc.

The whole course occupies three years, and forms an appropriate introduction to that kind of instruction which can only be obtained within the walls of the manufactory, or by actually taking part in the labors of the surveyor, the engineer, or the architect.

The following subjects are taught in this department :

1. Mathematics.

2. Natural Philosophy in its various branches, including Practical and Experimental Physics.

These are taught by lectures and illustrations in the usual way.

3. The Arts of Construction in connection with Civil Engineering and Architecture.

This consists of lectures on materials, foundations, the principles and practice of the design and construction of railways, bridges, houses, sewerage, tunnels, canals, docks, harbors, lighthouses, etc.; and the more advanced students are exercised upon essays on various engineering questions, and on constructive designs for works.

4. Manufacturing Art and Machinery.

This comprises lectures on the manufacture of iron and steel, and other metals, and on machinery and manufacturing processes of various kinds; the lectures "being intended to add a knowledge of practice to a knowledge of theory taught by the other professors."

The students have the opportunity of visiting works in the vicinity of London; and, at an advanced stage, are exercised on essays and designs as in the last-mentioned subject.

The instruction in this branch is aided by the establishment of an Engineering Workshop, in which the students are allowed to work, and where they have the opportunity of learning some of the simplest processes of practical working in wood and metals.

5. Land Surveying and Levelling.

These are taught theoretically by College lectures, and practically by exercise in the field.

6. Drawing.

This comprises not only free drawing, but geometrical projection, and practical drawing of the kind used by architects and civil and mechanical engineers.

It is taught by actual practice in the ordinary way.

7. Chemistry.

Taught by lectures and laboratory practice.

8. Geology and Mineralogy.

Taught by lectures and **by occasional** field excursions.

9. Photography.

Lectures and demonstrations.

Examinations are held, and prizes and certificates are given ; and a few of the students exhibiting most proficiency are elected " Associates," who are entitled to perpetual free admission, and to special honor in the College.

UNIVERSITY COLLEGE, LONDON.

This establishment is also designed to afford preparatory training to students of Engineering. It contains a *Department of Civil and Mechanical Engineering,* of which a special prospectus is issued annually.

The following extract from the prospectus will explain the general objects aimed at:

The course of instruction in this department is not intended to supersede the necessity of the

Engineering student serving a pupilage on the works of a Civil and Mechanical Engineer, as it is only upon them that he can obtain a thorough knowledge of the practical details of construction; but it is designed to teach him the theoretical principles of his profession, together with those habits of thought and observation without which he will not be able to take full advantage of the practice that will come before him during his term of pupilage.

The complete course extends over three sessions, and embraces the following subjects:

Mathematics, pure and applied.
Applied mechanics.
Physics.
Physical laboratory.
Chemistry.
Chemical laboratory.
Civil and mechanical engineering.
Mechanical drawing and designing.
Surveying and levelling.
Geology.

Students who have gone through the complete course, and who have passed the examination at the end of each of the three sessions, to the satisfaction of the professors, will be entitled to the General Certificate of Engineering.

There is also a Professorship of Architecture and Construction, in which these subjects are comprehensively treated in two courses, viz.:

Architecture as a Fine Art.
Architecture as a Science.

ROYAL SCHOOL OF MINES, LONDON

The object of this School is set fortn in the following extract from the prospectus:

"The principal object of the Institution is to discipline the students thoroughly in the principles of those sciences upon which the operations of the miner and metallurgist depend. Of course, nothing but experience in the mine and in the laboratory can confer the skill and tact requisite for the practical conduct of those operations; but, on the other hand, it is only by an acquaintance with scientific principles that the beginner can profit by that experience, and improve upon the processes of his predecessors."

The course of study occupies three years, and comprises the following subjects:

Physics and Applied Mechanics.
Chemistry (Inorganic).

Laboratory Practice.
Mechanical Drawing.
Geology.
Palæontology.
Mineralogy.
Metallurgy and Assaying.
Mining.

The mode of instruction is by systematic courses of lectures, by written and oral examinations, by practical teaching in the laboratories and drawing office, and also, under certain conditions, by field excursions.

Scholars who have gone through the proper course, and pass the requisite examinations, are entitled to receive an official certificate, conferring on them the title of " *Associate of the Royal School of Mines.*"

UNIVERSITY OF EDINBURGH.

The arrangements for engineering instruction at this establishment are thus described by the able Professor of Civil Engineering, Mr. Fleeming Jenkin, F.R.S.:

" The instruction provided for engineers at the University of Edinburgh is now, I think, fairly well organized. It consists of at least a two-years' course of study, arranged as follows :

First Year.

Mathematics.
Natural Philosophy.
Engineering.
Mechanical Drawing.

Second Year.

Summer Session.—Surveying, levelling, and setting out.
Winter Session.—Mathematics.
    Chemistry.
    Engineering.
    Mechanical Drawing.

These two years of study are followed each by a general University examination, leading to a degree of Bachelor, or Licentiate of Engineering. This degree has not yet been conferred, owing to cer-

tain legal difficulties, which are in process of removal.

The course called Engineering consists
of about 100 lectures in each session, and
treats of the following subjects:

Year A.

1. Principles of statics; couples; parallel and inclined forces; centre of gravity; moments of inertia.

2. Equilibrium and stability of frames,
bracing-cords, ribs and linear arches.

3. Strength of materials.

4. Strength of the simpler forms in
which materials are used.

5. Application of mechanical principles to combined structures, especially
roofs and bridges.

6. The construction of roads, railways
and tramways.

7. Principles of dynamics; Newton's
laws of motion; conservation and trans-
formation of energy.

8. Application of the principles of dynamics to prime movers, especially to locomotives.

9. Applications of machinery to manufactures.

Year B.

1. Application of statics to the determination of frictional stability and to hydrostatics.
2. Hydrodynamics.
3. The construction of waterworks.
4. Drainage of towns.
5. Construction of harbors.
6. Application of kinematics to machinery, illustrated by millwright work.
7. The construction of the condensing steam-engine.
8. The construction of water-wheels, turbines, and primary machinery.
9. Some special applications of machinery to manufacturing purposes.

GLASGOW UNIVERSITY.

In this University the Faculty of Arts, which contains Mathematics and Natural Philosophy, includes also a Professorship of Civil Engineering, the studies in which are summed up briefly as follows:

The stability of structures, the strength of materials, the principles of the action of machines, prime movers, whether driven by animal strength, water, wind, or the mechanical action of heat (as in the steam-engine), the principles of hydraulics, the mathematical principles of surveying and levelling, the engineering of earthwork, masonry, carpentry, structures in iron, roads, railways, bridges and viaducts, tunnels, canals, works of drainage and water-supply, river works, harbor works, and sea-coast works.

A certificate of " Proficiency in Engineering Science" is granted to students who have passed two sessions in the above studies, and also satisfy the examiners as to their knowledge of mathematics, natural philosophy, chemistry, geology, and mineralogy.

The following notes on the subject are by the present able Professor of Civil Engineering at this University:

Notes as to Instruction in Engineering Science, drawn up for the information of Students.

1. *Preliminary Education.*

Of the ordinary branches of elementary education *arithmetic* is of special importance to the student of engineering ; and he ought to be familiar in particular with the most rapid ways of performing calculations consistently with accuracy.

It is desirable that he should be well instructed in engineering and mechanical drawing, as part of his preliminary education ; but he may, if necessary, obtain that instruction during the intervals of a University course.

It is also desirable, *if possible*, that the elementary parts of mathematics, such as plane geometry, plane trigonometry, and algebra as far as quadratic equations, should form part of his preliminary education, as thereby time and labor will be saved during his University course.

2. *University Course.*

The course of study and examination adopted by the University of Glasgow is described in the Glasgow University Calendar.

In drawing up that course the University have had in view to avoid altogether any competition with the offices of civil engineers, or the workshops of mechanical engineers, or any interference with the usual practice of pupilage or apprenticeship; and they have accordingly adopted a system which is capable of working in harmony with that of pupilage or apprenticeship, by supplying the student with that scientific knowledge which he cannot well acquire in an office or workshop, and avoiding any pretension to give him that skill in the conduct of actual business which is to be gained by practice alone.

The University course may be gone through either before, during, or after the term of pupilage or apprenticeship, according to convenience. An arrangement which is sometimes found to answer well is to devote the winter to academic study and the summer to the practice of engineering. A student who is not a candidate for a certificate in engineering science may attend as few or as many classes as he thinks fit.

(Signed)   W. J. MACQUORN RANKINE.

OWENS COLLEGE, MANCHESTER.

A department of Civil and Mechanical Engineering has been added to this College.

The following extract from the prospectus will explain the course of education pursued:

The complete course of instruction in this department, extending over three years, embraces the following subjects:

First Year.

Mathematics.
Natural philosophy (mechanics).
Chemistry.
Geology.
Geometrical and mechanical drawing.

Second Year.

Mathematics.
Natural philosophy (physics).
. Chemistry.
Mechanical engineering.
Civil engineering.
Drawing and Surveying.

Third Year.

Mathematics.
Natural philosophy (mathematical).
Mineralogy.
Engineering (Senior Class).
Drawing and surveying.

Successful attendance on the course will furnish a thorough scientific groundwork for the
attainment of the knowledge requisite for the
prosecution of the higher branches of the engineering profession, but it is not intended to
supersede the practical training which can only
be obtained in the office of a Civil, or the workshop of a Mechanical, Engineer.

Certificates in Engineering will be granted by
the College. The examination of these certificates will comprise all the subjects recited
above.

GENERAL OBSERVATIONS ON THE POLY-
TECHNIC SCHOOLS OF THE CONTINENT.

The French commission remark on
these as follows :

POLYTECHNIC INSTITUTES.

"The various institutions intended
for commercial or industrial education
present, under identical designations,
very great diversities in Germany, but
the case is different with polytechnic establishments, which, under the name of
Gewerbs-Institut at Berlin, and of Polytechnic School or Institute in Saxony,
Bavaria, Austria, Würtemberg, Switzerland, and the Grand Duchy of Baden,
are intended to train civil engineers for
the services of bridges, roads, mines, and
manufactures, mechanical engineers,
manufacturing chemists, architects, forest engineers, etc. In all these establishments scientific instruction is given
in a very high degree, and sometimes
even to an extent superior to the re-

quirements and the end to be attained; but, everywhere also, the technical branch of this instruction is cultivated with the utmost care. The polytechnic institutes are at once schools of theory and of application, and present, in this respect, a very great analogy with the Central School of France.

" In all these establishments the pupils enter at seventeen or eighteen years of age, and must possess a preparatory education corresponding to the special studies they intend to follow. The choice of his branches of study having been made by the pupil, the courses he must attend are indicated to him, and become almost everywhere compulsory. However, this obligation is not always absolute, and the liberty accorded to the pupils, of not attending certain scientific courses, has the effect of inducing the professors to confine their theoretical instruction within the limits of what is really useful to those divisions.

"The part of the first courses which forms the scientific foundation of the technical applications is usually common to several of the special divisions into which the pupils are separated, and each division likewise receives the peculiar in-struction required for it. These divis-ions, more or less numerous according to the country, are in general the follow-ing :

"Engineers for bridges and roads.

"Civil engineers for railways, etc.

"Architects and builders.

"Mechanics.

"Manufacturing chemists.

"Mining engineers.

"Forest engineers.

"All the institutes do not comprise the same number of divisions, but the first four or five are almost universally adopted, if there be no special establish. ment to replace them.

"The peculiar arrangement and grad. ation of the studies nearly always possess a remarkable feature, which is that the first part of the studies of each special division, which requires one or two years, is so regulated that it constitutes a body of knowledge sufficiently complete to allow a young man to break off there and enter advantageously on the second-rate positions in the career he has chosen. After accomplishing this first part of the studies, a pupil may become an able assistant engineer of roads and bridges, or of civil architecture (*Werkmeister*), a builder (*Baumeister*), overlooker or head mechanician, a dispensing chemist, or foreman of chemical works, a head miner, a mining overseer, a forest agent, etc. In more than one state, pupils are even required, after reaching this first stage of technical instruction, to pass a year or two in building-yards, workshops, or factories, before going through the rest of their studies. This practice, which presents the inconvenience of interrupting the studies and exposing many pupils to the danger of forgetting a part of them, has, on the other hand, the advantage of maturing their minds by practice, of showing them the applications of science, and of not leading to higher studies any but those who really have a vocation for them. It is, however, practicable only under the system of out-door pupils, which is universal in Germany, and for pursuits in which there is no limit of age."

The duration of the instruction at the École Polytechnique is two years, that of the special instruction at the Écoles des Ponts et Chaussées et des Mines is three years. These three years comprise eighteen months of missions on works.

The preparatory instruction given at the École Polytechnique has comprised, from its foundation, the mathematical sciences, drawing, physics, chemistry, mechanics, hydraulics, etc., as well as the general sciences, astronomy, and geology. The pure science acquired in this school exceeded then, and exceeds much more now, the notions necessary for the special schools of Ponts et Chaussées and of Mines. But far from being disadvantageous, this was the germ of an intellectual development, most rapid and most elevated, for the choicest scholars, and to this must be attributed the large number of *savants* who have come from this school.

The admission to the Polytechnic School is by public competition, and to this liberal and democratic measure, as well as to the scientific success of some pupils, is due the popularity the school has enjoyed from the time of its foundation.

The Bachelor's degree in Science or Literature (*Baccalaureat ès sciences ou ès lettres*) is required for admission to the competition. The programme of the competition has undergone developments analogous to that of the instruction itself. It comprises now the whole of arithmetic, the elementary and part of the complete instruction of geometry, algebra, trigonometry and descriptive geometry, physics, and general chemistry.

But the separation of the two classes (military and civil) is no longer so absolute in the actual exercise of the profession as this classification would seem to indicate. A certain number of military engineers abandon the profession of arms in order to enter into an industrial career ; engineers of the Artillery and of the Navy give themselves up in the workshops and ship-yards of the Government to studies and labors by which industry profits. As the Government has demanded hitherto but little competition from private enterprise in the manufacture of the arms, the ships, and the engines of the Imperial Navy, it follows that important establishments are directed by State engineers, who seek to follow, if not to advance, the progress of private industry in works of the same kind.

The instruction in the special schools of the Ponts et Chaussées and Mines has, for its object, the application of the sciences to public works, to the working of mines, and to the treatment of mineral matters.

This range of scientific applications becomes more extensive every year ; it includes not only the public services, but nearly all large industrial enterprises.

The constant tendency of the State being to substitute its action for that of private industry in the public services, as well as in the industrial enterprises which are connected with any political or fiscal requirements, it will be seen how important is the basis of the education given by the State to its pupils, and what a multiplicity of details it involves.

In the terms of the special programme for the École des Ponts et Chaussées, the instruction comprises " the construction of roads, bridges, railways, canals, harbors, the improvement of rivers, civil architecture, applied mechanics, hydraulics, the steam engine, agricultural hydraulics, the geological and mineralogical knowledge required in the arts of construction, administrative law, and political economy."

The pupils are exercised in drawing operations, in the preparation of designs or projects, the manipulation and testing of materials, levelling, mechanical drawing, etc.

For half the year they are sent away to employ themselves, under the direction of the chiefs of the service, in the practice of the art of the engineer.

ÉCOLE CENTRAL DES ARTS ET
MANUFACTURES.

The studies of the Central School may be thus briefly recapitulated:—

In the first year pupils follow the course of descriptive geometry with applications; analysis, comprising the elements of the differential and integral calculus; kine-matics, general mechanics, general phy-sics, general chemistry, construction of machines, and hygienics.

In the second and third year courses of applied mechanics, construction and putting up of machines, analytical chem-istry, industrial and agricultural chemis-try, constructions (civil buildings, public works, and railways), applied physics and steam engines, metallurgy, mineralogy, geology, and working of mines.

The course of construction of ma-chines, which is very complete, as well as that of applied physics and steam engines, and the course of applied chemistry, are peculiar to the Central School. The teaching of mechanics is also conducted on a new plan, in a spirit essentially practical.

An accurate idea of the education given at the Central School may be formed from an attentive study of the new programmes. Whoever reads them will admit that if they have not yet reached perfection, they nevertheless present a well-ordered instruction useful to all who wish to pursue an industrial career.

The oral instruction of the Central School is judiciously completed by imposing on the pupils numerous studies of projects, by manipulations in the laboratory, by visiting workshops, by mineralogical and geological excursions, and especially by frequent compulsory examinations, not only at the end of each year's studies, but also during the courses and at their close.

HOLLAND.

In the Netherlands any one who pleases is perfectly at liberty to exercise the profession of engineer, architect, mechanical engineer, or engineer for the mines. Anybody, too, who chooses may style himself engineer, architect, etc.; but no *Government diploma* conferring such title is granted except to those who have passed the regular examinations required by law.

All engineers in the Government service, for the Ponts et Chaussées and the mines in India, must be duly qualified by the above-mentioned diploma.

Those who wish to obtain *a diploma* must follow a fixed course of study in order to prepare themselves for the examinations they will have to pass.

Government diplomas are given for Technology, Civil Engineer, Architect, Naval Engineer (ship-building), Mechanical Engineer, Engineer for the Mines (metallurgy).

These diplomas are granted on passing various examinatious, namely:

a. A general examination (called examination A) in the different branches taught at the Higher Burgher Schools, serving to prove the candidate's proficiency in all the preparatory studies.

b. Examinations in the engineering sciences themselves, differing according to the diploma demanded by the candidate: as for Civil engineer, two examinations are required; namely, an examination (called examination B) in

a. Higher algebra.

b. Spherical trigonometry and analytical geometry.

c. Descriptive geometry.

d. Differential and integral calculus.

e. Application of physics.

f. Analytical chemistry in relation to building materials.

g. The knowledge of building materials for architectural and hydraulic works.

h. The construction of the various parts of buildings.

i. Plain architectural and hydraulic as well as ordinary drawing.

An examination (called examination C) in

a. Theoretical and applied mechanics and the knowledge of machinery.

b. Hydraulic architecture, comprising—

1. The construction of roads, railways, and bridges.
2. Sea-dykes and embankments.
3. The knowledge of rivers as means of drainage and in relation to navigation.
4. The construction of canals, sluices, harbors, and maritime works.
5. The hydrography of the Netherlands, the knowledge of polders and mill-drainage.

c. Domestic architecture, comprising—

1. The construction of plain buildings.

2. The principals of ornamental architecture.

d. Topographical, ornamental, and ordinary drawing, as well as the drawing of objects required in hydraulic worsd.

e. The drawing up of plans and estimates.

f. The elements of geodesy and practical levelling and surveying.

g. Administrative law, relating to engineering and public works.

RUSSIA.

There exist in Russia five corps of Engineers of the State, viz.—

1. Military Engineers.

2. Naval Engineers.

3. Engineers of Maritime Construction.

4. Engineers of Ways of Communication.

5. Engineers of Mines.

These Engineers receive their education in the following establishments:

1. At the School of the Academy of Military Engineers.

2. At the School of the Engineers and Artillerists of the Marine.

3. At the Institute of the Engineers of Ways of Communication.

4. At the Institute of Mines.

All these establishments belong to the State ; there are no private ones of the kind in Russia.

In 1867 the Institute of Engineers of Ways of Communication, and in 1866 the Institute of Mines, were completely re-organized so as to furnish engineers, not only for the State, but also for private industry. Also, none of the students of these Institutions are now obliged to serve six years to the State, as was the case before 1864 for the students educated at the expense of the Government. The students of the Institute of Ways of Communication have changed their name to that of Civil Engineers, and the Corps of Engineers of Ways of Communication is now completed by these latter ;

the students of the Institute of **Mines**
have retained the name of **Engineers of
Mines.**

*On the Course of Studies followed at the
Institute of Ways of Communication.*

At the former, the following subjects
are taught:

General notions of theology.

Law, having particular reference **to**
constructions and ways of communication.

Political and statistical economy.

Chemistry, general and analytical.

Physics, general and technical.

Telegraphy.

Mineralogy and geognosy.

Mathematics, differential and integral
calculus, and analytical geometry of **two**
and three dimensions.

Descriptive geometry, and its application to the theory of shadows, to perspective, and to isometrical projection.

Drawing.

Topography, astronomy, and **geodesy.**

Statics and analytical mechanics.

Applied mechanics (kinematics, living prime movers, hydraulics and water conduits, aerodynamics), and the steam-engine.

Constructive mechanics (the theory of the strength of materials and of the stability of the parts of buildings).

The art of construction (studies of materials, works of construction and foundations, drawing and irrigation, the consolidation of lands and slopes ; the embankment of rivers, and other means of preservation against inundations ; sewers, bridges, roads, and railways ; the improvement of navigable streams, canals, harbors, and docks).

Architecture.

The preparation of designs for mechanchanical and architectural objects, for bridges and other constructions.

Every student is also obliged to study one of three languages, English, French, or German.

On the Nature of the Diplomas given.

The students of the fifth class of the

Institute of Ways of Communication, after having passed their final examination, receive a diploma, giving them authority for the direction of all kinds of works of construction, with the title of Civil Engineers.

PRUSSIA.

The Prussian system of combining Architecture and Engineering rules the plan of lectures at Berlin and Hanover. Mechanical engineers find in Berlin the "Gewerbe Academie," where extensive workshops offer to the students practical exercise, combined with theoretical instruction by lectures.

Those who devote themselves to the service of the State in Prussia have to pass through the following course of education.

1. General scientific education must be proved by a testimony of "maturity for university," which can be acquired by examination when leaving the upper class of a college.

2. Then follow three years' regular study at the " Bau-Akademie at Berlin," or the polytechnic school at Hanover. Part of this time may be spent at another polytechnic school.

3. One year's practice under the superintendence of State Engineers is afforded, but it is permitted to combine this with the academical years, by disposing the vacation time (several months every year) for practical purposes; so that by adding these together, and with a few months afterwards, a full year is to be made up. These three conditions being accomplished, the student may present himself for—

The First Examination.

Which comprises mathematics, including the differential calculus, all branches of engineering and architecture, theoretical and practical, geology and other auxiliary sciences, history of art, knowledge of styles, literature, perfect ability in drawing, architectural and engineering, versatility in projecting and estimating any kind of work, calculating and projecting machinery, as far as it is used for building and engineering purposes. The whole examination takes several days, by special examiners for the different studies. Each of these gives separately a certificate, viz. "distinguished," "good," "sufficient," "tolerable," "insufficient." These are brought in the full commission, which draws the result, "Admitted," or "Not admitted." In the latter case, the young man may present himself again in a year or two. If admitted, he receives the title of "Bauführer" (Conducteur), has to promise solemnly, correct, upright, and good behavior, is declared qualified for temporary paid employment under superintendence of an official engineer or architect, and has "public faith" in measuring and receiving work and materials, and also in carrying accounts.

Now he is at liberty to choose his way of further study and practice as he likes, and when he thinks to be sufficiently prepared he may apply for—

The Second Examination.

The form for that is to send in an application for " Aufgaben" (themes, problems) connected with the declaration that he wishes to be admitted; he is permitted to name that branch of the profession (engineer or architect) to which he has principally devoted himself. This will be considered, to a certain degree, but without dispensation from the other branch in general. In answer to the application, the candidate receives two themes, one in architecture, another in engineering, and two years' time is left to him for coming forward with his elaboration, which requires generally a year's hard work, consisting of about fifteen to twenty large sheets of drawings (plans, sections, facades, details of any kind, machinery, etc. etc.), and accompanied by a voluminous explanatory report. The board of examiners have then to consider whether the elaboration is acceptable, and if this is affirmed, a verbal examination takes place; the main object of this being to prove the true authorship of the candidate as regards the elaboration, and to test his general knowledge, judgment, scientific standing, and practical views.

BADEN.

At Carlsruhe in Baden there is a preparatory mathematical school, in which pupils are prepared for the

SCHOOL OF CIVIL ENGINEERING.

First Year's Course.

Strength of materials (Parts I. and II.).

Applied hydraulics and the mechanical theory of heat.

General administrative economy.

Political economy.

Freehand drawing. Landscape painting and water-coloring.

Hydraulic works (first course).

Road construction   do.

Constructive examples (first course).

Use and nature of mechanical instruments.

Construction of machinery (first course).

Chemical nomenclature.

Technical course of architecture.

Study of the orders of architecture (first course).

Exercises in architectural projects.

Constructions in masonry.

General history of art.

Second Year's Course.

General study of manufactures.

The general and more important special study of civil law.

Freehand drawing ; landscape drawing and water-coloring.

Hydraulic works (second course).

Road construction (second course).

The construction of railways.

Examples and exercises in the construction of hydraulic works and roads (second course).

Theoretical study of machinery.

Theory of the consumption of fuel and heating apparatus.

The manufacture of machines (Part II.).

Mechanical nomenclature.

Study of architectural orders (second course).

Exercises in architectural projects (Part II.).

General history of art.

Third Course (six months).

Hydraulic works (third course).

Road construction   do.

The above with special reference to the Grand Duchy of Baden.

Designing important engineering projects.

Marine works.

Practical geometry (Part II.).

The higher geodesy.

Method of fluxions.

N.B.—Every year excursions are made to examine works either in process of construction or already carried out.

For entrance to the Civil Engineering

Department of the Polytechnic School at Stuttgart, the student must as a rule have completed his eighteenth year.

He must possess a proper certificate of good moral training, and have acquired the necessary information.

When under eighteen years of age the consent of his parents or guardians is necessary to his entrance into the school.

He must give in a written declaration as to his education, and give evidence of possessing the knowledge without which he could not with advantage attend the professional courses of instruction.

He shall prove that he possesses this preliminary knowledge before the principal of the particular technical school.

The Engineering Course extends throughout three years.

First Year's Course.

Chemistry, mineralogy, and a knowledge of the structure of the earth.

Practical geometry : instruments for the measurement of angles ; plain triangulation, and fixing stations ; eradication

of errors ; trigonometrical and baromet-
rical levelling ; measuring distances ;
examples of foregoing ; trigonometrical
calculation of heights ; technical mechanics; stability of buildings ; elasticity and
strength of materials ; beam-girders,
arches, retaining and "revêtement"
walls ; solution of practical examples ;
statics and dynamics of fluids and gaseous
bodies, with special reference to their important applications in the practice of
Engineering.

Bridge construction (first course):
especially bridges in masonry, and retaining walls, with examples ; general structures in stone and timber.

Second Year's Course.

Bridge construction (second course):
with examples, viz., timber and iron
bridges; foundations; management during construction.

General study of buildings (second
course), with examples: iron and mixed
structures ; application of water-power
and steam ; examples in the construction

of machinery; history of the art of building (first course), especially with reference to the Grecian, Etruscan, and Roman manners of building.

Perspective drawing: drawing figures from plaster casts; landscape drawing from examples and from nature, in outline and shaded, in chalks, lead, pen and ink, brush and colors.

Third Year's Course.

Bridge construction (third course): waterworks; road and railway construction, with examples; staking out; calculating earthworks.

History of the art of building, with examples (second course): especially with reference to Roman and Gothic buildings.

AUSTRIA.

At the School of Engineering at Vienna the following is the outline of the course of study:

SCHOOL OF ENGINEERING.

First Year.

Building ; materials and the art of construction.

The mechanics of construction (first course), general division.

Mechanical technology.

Technical physics.

Analytical mechanics.

General knowledge of machines.

Ornament drawing.

Second Year.

Earthworks—setting out; tunnelling.

Construction of bridges.

Exercises in construction.

The Mechanics of construction (second course); theory of bridges.

Spherical astronomy.

Higher geodesy.

The knowledge of ground and Soils.

Third Year.

Road and railway construction.

Hydraulic construction.
Railway architecture.
Geology and mineralogy.
Law, as relating to buildings and Railways.

SWITZERLAND.

At Zurich the following studies are included in the engineering course:

ENGINEERING SCHOOL (three years' course).

First Year's Course.

Differential and integral calculus (with repetitions); descriptive geometry (with repetitions) and examples.
Study of the art of building.
Building drawings; plan drawing.

Second Year's Course.

Study of differential equations; differential and integral calculus (with repetitions).
Technical mechanics (and repetitions).
Plane geometry.

Study of shadows and perspective.
Topography.
Technical geology.
Technical physics (with repetitions).
Construction of machinery.

Third Year's Course.

Theoretical study of machinery.
Mechanics.
Theory of heat, and theory of the steam engine.
Construction of earthworks, bridges, and tunnels (with repetitions).
Examples in construction.
Construction of roads and canals.
Geodesy; Plan drawing.
Astronomy; administrative law.
Theory of construction of iron roofs.
The manufacture of rolled iron and the strength and calculations for iron girders.

In a School for Engineers at Rome, Italy, the following is the schedule of study for a three-years' course:

First Year.—Algebra, Geometry, Descriptive Geometry, Analytical Geometry, Physics, Principles of Architectural Designing.

Second Year.—Higher Algebra, Deteriminants, Calculus, Chemistry, Geometrical and Ornamental Designing, and Topographical Drawing.

Third Year.—Rational Mechanics, Graphic Statics, Geodesy, Mineralogy, Chemical Analysis, Geology, Construction of Machines, Bridges, Roads, and Hydraulic Works, Engineering and Architectural Designing, Mechanics applied to Structures, Properties fo Materials.

CHAPTER IV.

AT two important meetings of the engineering societies of this country the subject of discussion was Technical Education. The first meeting was that of the American Institute of Mining Engineers, at Washington, February, 1876, and the second a joint meeting of the Mining Engineers and the American Society of Civil Engineers, in Philadelphia, June, 1876.

A few abstracts of the addresses on these occasions are given below, with the names of the speakers.

The late Alexander S. Holley, president of the American Institute of Mining Engineers, in his address at the Washington meeting, said :

In order that the technical school should be in the highest degree useful, fruitful, and economical, it must instruct, not *men* of good general education, but *artisans* of good general education. The art must precede the science. The man must first feel the necessity, and know the directions of a larger knowledge, and then he will master it through and through. Mark how rapidly the more capable and ambitious of practical men advance in knowledge derivable from books, as compared with the progress of bookmen, either in books or in practice. Many men have acquired a more useful knowledge of chemistry, in the spare evenings of a year, than the average graduate has compassed during his whole course. These men realized that success was hanging on their better knowledge. Familiar with every changing look of objects and phenomena, they detected the constant play of the unknown forces which underlie them, and longed for a guide to their operation, as a mariner longs for a beacon light. This practical familiarity and judgment at once revealed the importance of scientific facts and methods, promoted their acquisition, and guided their application. Under what comparative facilities does the mere recitation-room student, or even the mere analyst of the hundred bottles, study applied chemistry? It is to these a matter of routine duty, without a soul; they are neither stimulated nor directed by a previously created want. Beginning with theoretical and abstract knowledge, is no less an inverted process in the useful arts than in the fine arts; as it would be to take a course of Ruskin within brick walls, as preparatory to opening a studio, and then climbing the mountains to square nature with the book.

Undoubtedly there may be extremes in any form of educational method. For a youth to begin the special business of technical education by any method, practical or otherwise, before he has acquired not only a common school education, but, at least, such a knowledge of polite literature and general science, including of course mathematics, as would fit him to enter one of the classical colleges, should be strongly discouraged, for various reasons. It is useless to disguise the fact that the want, not of high scholarship, but of liberal and general education, is to-day the greatest of all the embarrassments which the majority of engineering experts and managers encounter. This statement cannot be deemed uncomplimentary to the class, seeing that they have risen to power despite the embarrassment. At the present day, the high-school systems founded by states and by private enterprise, bring such an education within the reach of every one ; and it seems of the first importance to promote, if not almost to create, a public opinion, that liberal and general culture is as high an element of success in engineering as it is in any profession or calling.

But this is not all. Professional and business success is not, even in America, the chief end of life. All the social and political relations, and even personal happiness, are governed, not by the specialties, but by the balance of mental culture. What, then, shall we say of the policy of wealthy parents — not indeed general, but too frequent — of placing an uncultured boy in a technical school, and then in works and business, without giving him one chance to acquire a general and polite culture ?

Many young men display a liking, and others a marked talent, in some special direction. There is no danger that these will be crowded out of existence by the culture necessary to make a well-balanced mind ; and the nearer the talent approaches genius, the less imminent will be any such danger.

The proposition then is, not that mere common schoolboys shall go into works, and then into technical schools, but that young men of more advanced general culture, when they do begin the business of technical education, shall apply to nature first and to the schoolmaster afterward.

It may be urged in favor of beginning in the technical school, rather than in the works, that mental capacity for the

after acquisition and application of facts
and principles is thus developed. But
mental training is not the product of the
technical school alone. Habits of logical
thinking and power of analysis and gen-
eralization may be acquired in any school.
And a positive objection to beginning
with the technical school is, that it cannot stop at logical methods and sciences
which are essentially abstract. It also attempts to teach about objects and phenomena, the first knowledge of which, if
it is to be broad and genuine, must come
from the fountain-head.

These considerations may be further
illustrated by the course of the inexpert
graduate when he enters works as a matter of business or of study. We have
seen that the practical man can, at least,
keep the wheels running and the fires
burning, and that when he is of a certain grade of ability and ambition, he
will most rapidly acquire the scientific
knowledge and culture which, joined to
his practical judgment, make him a master. The unpractised graduate, however,

can keep neither wheels turning nor fires burning; he has not even the capacity of a conservator. Nor can he for a long time recognize, in the whirl and heat of full-sized practice, the course and movement of those forces about which his abstract knowledge may be profound. The youngest apprentices are more useful in an emergency. He must begin with the lowest manual processes, not indeed to become simply dexterous, but, as it were, to learn the alphabet of a new language. He has started in the middle of his course instead of at the beginning. He must go back before he can advance, while the practician goes straight on. The knowledge of the schoolman about physical science, however often he may have visited works and mines and engines during school excursions, is essentially abstract; it no more stimulates desire and power of practical research than the calculus creates a passion or a capacity for studying the actual work of steam in an engine, or the actual endurance of a truss in a bridge.

The disappointment of inexpert graduates at finding themselves so far from being experts, their inability ofttimes to pay for further schooling, the necessity that they should now begin to earn money, as they had persuaded themselves they could so readily do upon graduation, discourage many from pursuing engineering, and, what is worse, send many out into practice who never do complete their technical education, but who, by the character of their work, lower the professional standard.

It can hardly be urged against the precedence of practical culture, that the student will get "out of practice" while he is in the school. He may, indeed, lose dexterity, but not the better fruits of experience. In fact, those who begin as practicians, almost instinctively keep up their intimacy with the current practice.

A most signal advantage of beginning technical education in the works is, that the mind is brought into early and intimate consideration of those great elements of success which cannot be imparted in any other way—the management of labor and the general principles of economy in construction, maintenance, and working. *An early knowledge of these subjects moulds the whole character of subsequent education and practice.* There seems to be no corresponding advantage in beginning with the technical school. The fundamental mathematics and general information on physical science may be acquired in the preliminary school.

There is little doubt that the managers of technical schools will favor this order of study. They want to graduate, not half-educated men, but experts. They desire, of all qualifications in the student, that enthusiasm which can only spring from a well-defined want of specific knowledge.

2d. But the *order* of education is not the only desirable change. Whether before or after their course in the school, the hundreds of young men who are every year entering engineering pursuits, are wasting their time in bad methods of practical study, or, if after the school course, they are more frequently doing bad work as engineers, when they should still be only students. Hardly two engineers acquire any part of their practical knowledge in the same curriculum. They pick it up as best they may, usually in a manner that is wasteful of time or damaging to the public. While the teaching of general facts and principles and of scientific method is highly developed, there is no organized system for guiding students to direct knowledge of objects and phenomena. This statement requires two explanations : I. Apprenticeship is a school of skill in a specialty rather than a school of liberal art. It is intended for a class of men who propose to remain mere workmen, and not for the class who intend to improve and direct engineering enterprises. It imparts a degree of dexterity far beyond the requirements of the general expert, while it would hardly impart in a lifetime his required range of practical knowledge. II. A school of engineering practice, such as that of research in zoology which was established by Agassiz, would be wholly impracticable, because it could be nothing less than a vast and successful establishment for construction and operation in nearly all the departments of engineering. If such a school were not commercially successful, and if its range were not comprehensive, it would be unsuitable and inadequate.

Now, if there can be a *system* of instruction in the one school, there can be in the other. The same discipline and responsibility, the same guidance as to precedence of study, quality of evidence, and correctness of conclusion, should hold good in both cases. To say otherwise would be to say that *all* knowledge should come from unaided original research, and that every investigator should begin, not where a former investigator left off, but where he began. It therefore appears that there can be a school of practical engineering, but that it cannot be mere apprenticeship in engineering practice, nor a system of engineering construction and operation, maintained merely for the purposes of a school.

The only alternative is to establish organized schools in the various existing engineering works. At first, this ·idea would seem subversive of all discipline and economy, but I am assured by experts in several branches of engineering that such would not be the case. Let us take, for example, a Bessemer works. A score of students under the discipline, as well as under the technical guidance of a master, could be distributed among its various departments, not only without detriment, but with some immediate advantage to the owner, for while receiving no pay, they would become skilful, at least as soon as the common laborers who form the usual reinforcements. Students should, of course, be expected, not to work when and in what manner they might choose, but to do good and full work during specific hours. This responsibility as workmen would rapidly impart not only the knowledge sought in the works, but a desire for higher knowledge and culture.

These considerations are not merely theoretical. Several students at a time, subjected to no discipline, sometimes working hard, and sometimes not at all, may often be found in a Bessemer works, and I have yet to hear of their embarrassing the management in any way. The laborer has no cause for interference, as the students are not under pay, and whatever they accomplish is clear gain to the three parties concerned—the owner, the student, and the operative. A large number of young men may be found studying in machine shops, and sometimes earning small pay, besides having opportunity to work in all departments.

The proposition is to enlarge and systematize the existing desultory study in works—to increase its usefulness to the student, and, at the same time, to make the granting of such facilities to students an object, immediately, as well as remotely, to the owners of works. To this end, the schoolmaster should be not only well read in the professional literature, but a practical expert who could take charge of the works himself, so that whilst best aiding the students, he could prevent their interference with the regular and economical operations. His functions would be, not those of an instructor, nor, to any great extent, of a clinical lecturer, but those of a disciplinarian. The students should acquire skill, in order that they might acquire judgment of skill and original knowledge of materials and forces, and the master should see that they did acquire them all. He might do some service by stated examination and current criticism and suggestion, but his chief office would be to promote honest work, and to provide opportunity for work in all departments with reference to the economy of the student's time and to the owner's interests.

It should thus appear that these somewhat radical changes in the curriculum of engineering study—first, a hand-to-hand knowledge, acquired not desultorily, but by an organized system, and afterwards the investigation of abstract and general facts and their relations, would largely economize the student's time and better the quality of his knowledge. The novice is nearly as valuable a student in works as the graduate, but he is a vastly less apt scholar in the school. My own belief, founded on the study of many typical cases, is, that this order of procedure would produce a better class of experts in little more than half the time required by the reverse order; that it would always make *experts ;* that it would discourage none from finishing an engineering education which would be complete in its parts, even if insufficient time were taken to fully develop it. A well-balanced culture will naturally grow in scope and in fruitfulness.

In this connection it seems proper to say a word about the royal road to learning, which a few ill-advised students attempt to pursue. I do not refer to their availing themselves of professional data and drawings on file in engineering offices, but I do refer to their asking engineers and managers to furnish them special reports on subjects regarding which their own observation would be vastly more useful to the applicants, and quite as convenient to the respondents — reports on the number and duties of workmen in each department, and the particulars of operation and relative cost, which can only be profitably investigated by a student, when not only the facts but the reasons are ferretted out by himself, rather than transmitted to the academic grove through the post-office.

In conclusion, if it should appear upon larger observation, to the profession in general, as it does appear to many of its members, that this want of coalescence. ranging from indifference to antagonism between its scientific and practical branches, is a real and substantial fact, a larger effort would undoubtedly be made to change a condition so damaging to the profession and to the public. This inappreciation of one department by the other is not unnatural—neither side has taken sufficient pains to observe what the other side has done. The mere scientist instinctively believes that the achievements of the profession are so far due to the deductions of scientists that all other causes fade into insignificance; and the practician knows that just as far as animal life is from the disembodied spirit, so far is utilization of nature from the formulæ of heat, chemical affinity and mathematics itself.

The first step is to recognize the fact, and I beg engineers, especially those who, from their scholastic habits, see least of the everyday embarrassments which are encountered by the executive departments of the profession, to take into account, not only the pride of class power, which the artisan feels as keenly as the scientist, but those baser elements of disunion, ranging from trades-unionism to counting-room dictation in technical affairs.

Having recognized the grave and comprehensive character of the evil, the next step should be, not I think, to attempt any violent alteration in the existing conduct of engineering by the men who are now in active service, but to change, if I may so say, the environment of the young men who are so soon to take our places, in order that their development may be larger, higher and in better balance. Two co-operative methods have been suggested—reversing the order of study, and organizing the practical school.

Whatever the course of improvement may be, it becomes us to leave some heritage of unity to the coming race. How shall we more fitly crown a century of engineering—a century in which our noble profession has risen from comparative potentiality to living energy ? And as its force is multiplied by the general advance of science, it becomes the momentum which evermore shall actuate the enginery of civilization.

Prof. Robert H. Thurston at the same meeting said: The question to be discussed here is certainly not whether the young engineer shall have a technical training, but whether he shall secure it in one way or another of several proposed methods. We are not asked whether he shall have such an education and training, but how shall we give it, and when should he seek it, and where.

I have said that I would specify three courses, either of which may, perhaps, accomplish the desired result. These are:

First. That method which is most usually adopted, in which the student is given his education, and is then sent into business.

Secondly. That which gives the boy a common-school education, then sends him into the office, or the field, or the workshop, to acquire a certain amount of practical experience, business knowledge and general development, and finally places him in the technical school, to obtain the professional education and scientific basis for a sound reputation, which can there be best and most readily given him.

Thirdly. The course which, although usually most difficult to pursue, is, if I may judge from observation and expe-

rience with a considerable number of instances, the most perfectly and econom-
ically successful. That is, a mixed course
of study and practice, extending throughout the early life of the man up to his
final and complete immersion in the
practice of his profession.

It is possible that I may be influenced
by that prejudice, which most men have
in favor of a course which has answered
its purpose more or less fully in their own
cases, or in cases which have appeared to
them illustrations of great or of even
moderate success; but I believe that the
boy who, with natural predisposition
toward a certain branch of engineering,
spends his weekly holidays and his vacations playing about the workshop, grow-
ing up in contact with the workmen, and
witnessing continually all those operations which, as he becomes old enough,
he learns to conduct himself, imbibing,
with that wonderful accuracy and rapid-
ity for which boys are remarkable, all
the traditions and recognized principles
of shop practice, learning the construc-

tion and use of tools, and now and then acquiring the art of manipulating a machine or handling a tool, I believe that this boy will most easily and perfectly secure the technics of his profession.

The course adopted generally in this and in all countries is the first of those specified. The boy is sent to school, and is given the usual common-school education. Upon concluding this course of study, he is sent to the college, or the technical school, and a four years' course of higher education having been completed, he is sent into business at the age of twenty or thereabouts.

He has then been engaged in the work of the student all his life; habits of study have been formed, and usually he has become, to a certain extent, unfitted for the vastly different kind of occupation which is now to be taken up. He has acquired habits of study, a good memory, and the ability to utilize it thoroughly, and has learned to make logically correct deductions from properly grouped facts. He lacks usually, how ever, the power of quickly perceiving and promptly acting upon such perception. He probably lacks decision, has lost some of that strength of character which may have been his by inheritance, and he has none of that experience which is as essential as character and knowledge to success in business. He may possess a great store of learning, both general and professional, a well-trained mind, a sound judgment, and all that scholastic habits and training can give him, but he lacks the no less essential knewledge of men and of things which he can only obtain by a personal contact. He cannot manage his employés without either making unreasonable demands upon them or yielding to them more than is just. He knows nothing of methods of conducting business, and cannot have become accustomed to the hard rubs which so seriously disturb the tyro, and which so often discourage him at the outset.

Habits acquired in youth are always difficult to modify in later years. His **habits** are those of the student, and he must inevitably find it a seriously diffi-cult matter to acquire the peculiar and distinctive habits of the business man. Once succeeding, however, he will rarely fail of full success.

The exceptional course in this country, and I presume in Europe, is the second of those outlined. The boy goes to the shop, or the office, immediately after completing his grammar or high school course, and learns the trade which leads most directly toward the profession which he proposes to enter; or, under the tuition of some practitioner, he acquires a knowledge of the ordinary routine of work and some idea of the character of the greater problem which he may expect to be confronted with in later years.

Arriving at the age of twenty, he sees the advantage of the possession of a knowledge of the science of his profession and he leaves his practice for a time and devotes himself to study in some technical school.

His difficulty now is to acquire habits of study and the student's power of making his own that knowledge which he finds in books, and of grasping experimental data and of collating essential facts and grouping them systematically, and of deducing from them general laws of precise definition, and of well-determined range of application. He has to reacquire the mathematician's power of basing upon a statement of accurately defined conditions generalizations which find practical application in every department of human knowledge. He has to regain that fondness for research and study of which his business life has done so much to deprive him.

On the other hand, he has learned by experience to prize knowledge, both for its own sake and for what it will enable him to accomplish. He has learned in what direction he is to expect most aid from literary attainments and scientific knowledge. He can, to a certain extent, distinguish between those branches of study which give only a mental gymnastic training and those which enable him to accomplish two objects simultaneously

—to acquire a store of valuable knowledge and, at the same time, to profit by a no less useful mental fulness of stature; becoming more of a sage and more of a man at the same time.

On the whole, I suspect that the advantages of this method more than counterbalance the disadvantages, and I have no doubt that, could this course be generally adopted, it would be seen to have an importance in the acceleration of professional progress which we probably hardly realize to-day.

Two great obstacles intervene to prevent the general adoption of this plan. The first is that conservatism which always retards the introduction of anything new, and which usually makes it necessary to agitate for at least a generation before a really great change can be brought about. Even when all are agreed on the question of the propriety of a step such as this, it is usually a long time before the public inertia is fully overcome. This is well illustrated by the fact that the necessity of technical education itself—proposed two centuries ago, and fairly inaugurated a century ago by Vaucanson, the father of the great *Conservatoire des Arts et Metiers* at Paris—is only just now beginning to be universally acknowledged by even those who are not hampered by a traditional proclivity in favor of the old Greek non-utilitarian idea of a purely gymnastic system of education.

The second great obstacle is the natural, and almost universally observed, reluctance of the young man, who has once become fairly inducted into business, and who sees opportunities opening to him in the immediate future, to give up all and to return to the school to secure advantages which his reason tells him are still more important, but which he, nevertheless, cannot fully realize, looking upon them as he does from a standpoint which does not permit him to see them as distinctly as he may in after life, when experience has confirmed the previous judgment. An active, energetic, and ambitious young man can rarely bring himself to the point of going back to the school after having once tasted the pleasures of success in business. Where this has been done, however, it has been almost invariably the fact, if I may judge from my own observation of quite a number of cases, the result is a most encouraging one. Could this plan be generally adopted, it would not only be decidedly better for the young man himself, but it would prove vastly better for the schools.

The greatest difficulty met with in carrying out a satisfactory course of technical instruction in the schools, is that of finding students who have sufficient ripeness of intellect and of judgment, and sufficient physical strength to comprehend readily, grasp fully, and retain perfectly, the principles which are presented to them. Boys are sent to technical schools without well-developed habits of study, with insufficient and superficial preparation, with minds unripe and with bodies still taxing their systems by the drain of that vital power needed in carrying on the operations of physical development. Were the la.t considered plan adopted, they would come to this work, which demands all the powers of maturity, with body and mind full developed, and with an understanding of the extent, difficulty, and importance of the work to be done which would insure vastly better performance, and the accomplishment of vastly more in the time assigned to the course. The work of the instructor would be rendered more easy and more satisfactory to both himself and his pupil. The time would be far better utilized, and the greatest good would be accomplished in the given time and by the expenditure of the given amount of time and funds.

It is in this direction, I am pleased to find, that our President is looking for higher efficiency in technical training. If the plan which he proposes, of making our larger manufacturing establishments advanced technical schools, can be carried out, it will prove, I am sure, a long step in the right direction. The

final portion of the work of education
would be done at a time when the
student has attained sufficient maturity
to appreciate it, and in the midst of such
influences as will most effectually impress its value upon him. I sincerely
hope that a way may be found of initiat-
ing this method of tuition, and that we
may soon learn just what we are to expect from it. Difficulties will undoubtedly arise, but with the exercise of care
in the choice, from among the many of
the few who are adapted by nature and
inclination to the pursuit of the profession, and with tact on the part of the
instructor and a hearty good will on the
side of the manager of the works, both,
hand in hand, working for the accomplishment of an object the importance of
which both appreciate, there can arise
no insurmountable obstacle to final and
complete success.

During the general discussion Mr. Os-
wald J. Heinrich remarked as follows:

There cannot be the least doubt, that
at no time in the past has a partial edu-

cation and training been sufficient for a man to fill, creditably, any position of importance in the various callings of the technical profession, and we need not expect that it will be different in the future. While partial experience alone, after perhaps a lifetime spent in a particular calling, may fit a man for a specific purpose, facts are not wanting to show that even such a man may commit great errors, or even blunders, in disregarding well-known principles which would be thoroughly understood by one of far less practical experience, but possessing a school education. It is true, a really good practical man, with indomitable energy, may succeed ultimately, but probably only after dearly-bought experience, which otherwise could have been avoided, and wasted time and money saved. This being the era when the state of cultivation of a nation is measured, to a great extent, by the most thorough use made of waste material, we we may just as well say, also, that this should be extended to the imponderable items of time and brain. On the other hand, a young man of a thorough theoretical education may deliver a lecture before a set of practical men which would fill them with admiration and awe, and yet he might be puzzled by the same set of men if called on to show how to do some simple practical operation. But the result may be, that other practical men may exult over the apparent superiority of the " indomitable " spirit of the practical man (probably themselves too ignorant to judge of the cost of his experience), and sneer at the failure of the scholar, and thus bring discredit on the attempts at liberal education of mankind. A long life in various practical callings may fit a man to fill even eminent positions, and, by being cautious, he may avoid such losses as have been enumerated, while, on the contrary, the unpractical scholar, possessed, as it is frequently the case, of too much self-reliance and mere book experience, may waste time and money to overcome practical difficulties. It is therefore not surprising to see the scale often over-weighted on the practical side.

It follows naturally, particularly in this country, that preference is given to the practical man. Unfortunately, for want of thorough understanding of the subject, the choice often falls upon a so-called practical man, and educational training has fallen into disrepute. The great drawback to obtaining a thorough, practical, and theoretical education in this country, will probably be less found in the means offered, than in the unwillingness to spend the time and money necessary to obtain it.

The order of the day in this country, "to make money," and, to a great extent, judging the capacity of a man according to the amount of money he has, or is earning, will unquestionably be a great drawback yet awhile. On the other hand, it is also impossible to obtain sufficient knowledge and experience during a few years of training, and in one particular course of instruction.

Taking my own experience, I had the

good luck, from my early boyhood until I had arrived beyond the years of maturity, to be alternately occupied in practical pursuits, and in receiving educational training, at schools of various grades. Until I had arrived at the age of twenty-three years I had never earned money worth speaking of, but spent my time from my fifteenth year in apprenticeships and going to various technical schools and public works. In my country, boys intending to devote themselves to technical occupations generally pursue the following plan, partially even regulated by law. After passing through the higher grades of the common school up to fifteen or sixteen years of age, where even, to some extent, Latin and Greek, but particularly modern languages, and the elements of mathematics and natural sciences form a part of the system of instruction, they are regularly apprenticed to the particular branch of business they intend to take up afterwards. As apprentices, they pass their regular time as carpenters, masons, pattern-makers, moulders, machinists at mines or furnaces, etc. Generally night schools, or schools during part of the winter—industrial schools—are visited during their time of apprenticeship, the time so spent being allowed as regular apprenticeship. They receive little or no pay during this time, according to choice snd circumstances. After spending several years in this way, they enter the higher grades of the technical schools or colleges, to pass through a thorough course of scientific training, at the same time, in various ways, being constantly reminded of the practical duties necessary to be performed by them hereafter, by making excursions during the period of lectures, and during vacation visiting the public works and shops of the country. After graduating at these schools, they enter again for a time as volunteers at the different public works or private establishments, and are glad to be taken as such, without receiving any compensation, sometimes even paying for the privilege. After such a course, and proper examinations, they are only considered, even at private works, to be fit to take a subordinate position, and are often only too glad to get it.

I consider a good general education more than desirable before entering practical life for various reasons. The principal reason is, that the mind of the boy is more susceptible to mental training and exercise. During his apprenticeship, or attention to practical work, he will find out the great help he may derive from educational training. This is kept up by attending the night or industrial schools during that time. These preliminary studies, connected with practical exercise, will balance mind and body, both essential for a young man in those years of life. He will be by far a more attentive scholar at the higher grade schools, at least so far as my experience has gone, and will profit more by attending such schools than generally is the case with those who have first passed through the entire collegiate or classical course of studies. I consider this a very

natural consequence of the necessary
course of studies in industrial schools,
they being better designed to prepare for
subsequent training than the old faculty
studies of law, medicine, theology, etc.

At the joint meeting of the Mining
Engineers and the American Society of
Civil Engineers, in Philadelphia, in discussion of the subject of engineering
education, Dr. R. W. Raymond, the
President, made an address of which the
following is an abstract: "I wish to
emphasize what I remarked on a former
occasion, that whether technical instruction be preceded or followed by manual
practice, one thing must precede both,
to insure the highest success in any
profession, and that is general culture.
For success is a social matter; it depends upon a man's influence over men.
Knowledge of facts and laws in nature
will not achieve it. The most thorough
metallurgist or engineer needs to be able
to make other men recognize his ability.
Nay, long before he can acquire thor-
oughness, he is dependent upon other

men for every chance of practice. A liberal education gives power over men; and the technical education, which gives power over matter, will be twice as easily gained, and twice as effective when gained, if it is grounded upon the mental discipline and the moral strength of a culture wider than its own.

The more one observes the careers of men about him, and the more one wrestles with difficulties of one's own, the more profound becomes the conviction that a young man makes a great mistake, who, because he is going to take a technical education in engineering, deliberately decides that he will not have any general culture to begin on. I am not speaking of the men who, struggling against cruel necessity, make their way honorably and effectually, in spite of early disadvantages. Such often win a place among the greatest names. But the reason is very simple. It is just the same reason as makes the Indian a hardy son of the forest. Excessive exposure, hardship, insufficient food and clothing,

do not make men hardy; they merely
kill off the men who are not hardy, and
those who survive must be the vigorous
ones. Poverty, ignorance, isolation, difficulty, are not elements of strength;
they are obstacles over which strength,
and strength only, can triumph. Infi-
nitely better they are than the luxury
that drowns ambition and breeds swamp-
gases of indolence and vice; but in themselves they are hindrances. A man who is
truly a man will not be enervated, but enlarged and stimulated by liberal culture.

I would appeal to no one sooner than
to our self-made men for a hearty recog-
nition of the value of such preparation.
They have felt the lack of it too keenly
not to wish for their children a better
chance. Now, with due caution against
the waste of time, I cannot doubt that a
general culture, though it may not be
the quickest preparation, will lead to the
best results. I remember the remark of
a man of great success and quick observation, who assured me that if his son
would become a metallurgical engineer,

he would put him through college first, and let him begin his special studies afterwards. I am not prepared to say that an entire college course is necessary, or that it is the best preliminary course, though I have a high opinion of it, but something equivalent to it, or to a part of it, that is, what our German cousins give to their young men in the *Gymnasium*. They give to them a liberal culture in the beginning; and it is a very remarkable thing, that amongst the many skilful metallurgists and mining engineers from Germany with whom I have had the pleasure of becoming acquainted, I have found a large proportion who had learned Greek and Latin, could perhaps even play on some musical instrument, and were widely acquainted with literature.

Finally, we must recognize the fact that individual character is, after all, the decisive element in success. We may devise plans without end to facilitate the manufacture of skilful engineers, but the men who have fidelity,

honor, virtue, courage, and that genius
which has been well defined as the power
of application, will make their way surely to the top, either by the help of our
arrangements, or in spite of them all ;
and of these born and bred leaders of the
profession, those who have the broadest
culture, other things being equal, will
stand easily first.

ABSTRACT OF MR. THOMAS C. CLARKE'S*
ADDRESS.

I would recommend that the engineering pupil get as sound a general education as possible, including the principles
of the sciences. Let his early education be
rather that of general culture, developing
his mind, strengthening his powers of
observation and judgment, teaching him
to generalize. This course he should, if
possible, pursue up to the age of eighteen
or twenty. Before that age the mind
and body are not generally sufficiently
developed to endure the physical hardships of engineering. Then let him

* American Society of Civil Engineers.

spend several years in practice in the machine shops, in the field, in the drafting-room, and in the office. Let him learn to deal with men and things, and to understand the conduct of affairs. Whether he will return to his books again depends upon what sort of a man he is.

I believe that all men, or nearly all men, from the natural constitution of their minds, fall into one of two classes.

They are either the men of executive ability, the practical men, *par excellence*, those who have a natural talent for affairs, the organization of labor, and the direction of men ; or else they are the men of science, the investigators, the men who are hungry for knowledge, and will learn the reason why. Very rarely one man unites both qualifications. James Watt did, and so did Professor Morse, but such men are rare.

If the young engineer belongs to the executive class, having once plunged into practice, he will probably never go back to his books. But the other kind of man will do so, either by himself or in the schools. When he will have found out exactly what his deficiencies are, and he will be able to judge (much better in most cases than his professor can) what it is desirable for him to learn, you may be sure of one thing, he will study the principles of science, and pay very little attention to their application as taught in the schools. He will not spend his time over the pages of Rankine, learning how English permanent way was made twenty years ago, before Mr. Bessemer was heard of. Whatever he studies will be of value to him, and no one can judge of that better than he can.

One thought more and I have done. To all classes of engineering students let me point out the immense value of acquiring and fully understanding the *scientific method*.

This is, first, the art (for it does not come by nature) of observing facts and acquiring data; second, of observing the relations of phenomena and of drawing conclusions therefrom; third, of verifying those conclusions by observation and experiment.

Robert Stephenson, in alluding once to the vast progress of modern engineering, in which he himself had born so distinguished a part, said : " We found it a craft, and we have left it a profession." That is to say, it had been put on a scientific basis, and by the use of the *scientific method*. This, after having been applied to the construction of railways, is now beginning to be applied to their management, and the results are remarkable, and promise to be more so.

ABSTRACT OF THE ADDRESS OF MR. COLE-
MAN SELLERS, AM. SOC. C. E.

It is safe to say, that a young man, after passing through college properly, and having a good sound education, who determines to succeed in the workshops at any hazard, will in two years make himself so valuable in the position that he occupies, as to be elevated by his employer into something higher.

Now I say that I thought of this very

deeply in the case of my own sons, and I did precisely, and I am doing precisely what I have just told you. I was not at all surprised, when I found my eldest son, after leaving the university, accepting a position in the workshop a little better than a common laborer. He commenced by chipping the scale out of the boiler. I tell you it was the best thing for him, because he made a beginning at the bottom, and did not shirk his work; it was as much as to say that he was willing to learn all that could be taught him in the shop, and he rapidly rose to a position higher than many who had been longer at work, but who had less book learning to back them.

It is impossible to make engineers out of pupils who have not engineering ability. There must be something in them that will compel them to take it up as a profession, and succeed in it. I am now clearly of the opinion that as it is not in the power of most young men to take the college course, and then afterward to take the technical course; that it is far better for them to obtain what scientific knowledge they can in a good college, or in a technical college where something else is taught besides the exact sciences, where they can be taught the languages, not the dead languages but the modern languages, and taught at the same time rhetoric, composition, and all that will enable them to express themselves ; and by all means let them have a good sound basis of mathematics before they venture their education in the workshops. Then when they have entered the workshops there will be time to acquire technical education without schools. I have no doubt that many who have been liberally educated, have, after entering the shop, felt the want of some technical education, and have broken away from the shop, and gone into schools to learn. They felt the need of obtaining more knowledge, and that the time they spent in the college or school was not sufficient.

I do not think it advisable as a rule however, to take the boy from the work-

bench and send him to school a second
time. I have in some instances noted
the effect of such a course upon young
men to be disadvantageous. If the boy
has left school too soon, and feels afterwards the want of more knowledge, it is
well enough, if he can, to return to his
studies, but such return makes sometimes
a disadvantageous break in his habits. I
look upon it rather as a means of mend-
ing a defect in education rather than a
course to be pursued as prearranged with
an object. By not attempting to teach
too much " practice " in the schools, time
is left to give a good grounding in gen-
eralities, which cannot fail to be of use
in any walk of life, and which can be
better acquired when one is young. The
practising engineer has not only to master
his profession, but he must learn how to
place himself and his works before men
so as to be seen of them and appreciated
by them. He requires a very extended
knowledge ; all learning will at one time
or another be of use to him ; and habits
of study, which will enable him to con-

tinue a student to the end of his days, will the more readily fit him to rise in his profession, and make him a leader among men.

ABSTRACT OF THE ADDRESS OF COL. W. MILNOR ROBERTS, AM. SOC. C. E.

It may be necessary, or at least advisable, when considering the subject of the proper method of training the young engineer, to have special reference to the particular branch of engineering he intends to follow. For any branch there must of course be a proper foundation, to the extent of a good English education (if German and French are added, it would be decidedly advantageous), and a ready use of figures, and of mathematical principles, to precede both technical engineering study, and practice—this in any branch.

In mining engineering particularly, the student, to be reasonably accomplished, should also understand chemistry, as well as geology and mineralogy.

In the other branches of engineering, chemistry may not be so necessary or important, although it is a kind of knowledge which is useful to all engineers. An accomplished civil engineer should be familiar with mechanical engineering, and not ignorant of mining engineering, though he need not, necessarily, be an expert therein ; it could hardly be expected of him. His chief or highest duties are not embraced in either of those branches, and his principal requisite is ready, sound judgment, and the more this is strengthened and confirmed by experience, the better for his employers as well as for himself. Sound judgment can never be wholly the result of education, either technically in the schools, or in engineering practice, because it does not always accompany knowledge or even experience. For civil engineering, the teaching and training in those higher schools, where this department, with the use of instruments, is a regular course, the student can learn all that is necessary for him to know, before taking a very subordinate position in a regular engineer corps in the field, where he would still have much, very much, to learn, which cannot be conveyed to him thoroughly in any other than this final school.

It may not be equally practicable to organize "practical schools under the direction and discipline of experts in engineering works," in all of the branches of engineering, but in mining and metallurgical engineering it seems to me to be quite practicable and desirable, likewise in mechanical engineering. In civil engineering the real school is largely in the field, beginning with the rapid preliminary explorations of lines of canals, or railroads, or projections of water-works, etc., extending through the processes of provisional and final locations, up to the planning and construction of the various works and structures appropriate to the particular improvement. Those who by great experience become experts in civil engineering, are usually too closely occupied in the professional conduct of works to take an active or controlling part in the business of educating younger members, excepting as above indicated, by having them in their corps on active duty of some kind. The day may come in this country when civil engineering may assume a somewhat different shape, but at present it appears to me that the polytechnic schools in our country, in which civil engineering is a leading feature, furnish adequate training for young men desirous of becoming civil engineers. Of course the more thoroughly the teachers are themselves grounded in the practice as well as the principles of civil engineering, the better it is for the pupils, though it may be well to consider that the most expert, and the most experienced in practical engineering, are not necessarily the best teachers. There are men peculiarly well adapted to shine and succeed as teachers of young engineers, who would not be selected to take the responsible practical management in particular lines of civil engineering, while there are many instances of good practical engineers who would be likely to do no honor to a technical professorship.

FROM THE ADDRESS OF MR. ASHBEL WELCH, MEMBER OF THE AMERICAN SOCIETY OF CIVIL ENGINEERS.

I suppose all agree that the future engineer should remain in the school or college till he is eighteen or twenty years old, and should get all the general education he can, up to that time, before he begins his professional education.

But experience shows that a *long* course of technical study, preceding and unaccompanied by professional practice, is highly inexpedient. I propose to glance at some reasons why it is so.

The object of the philosopher is to attain scientific results; the object of the engineer is to attain directly beneficial ends by using those results. One gets up the tools, the other works with them. Engineering education should therefore aim at readiness and skill in the application of science, rather than at scientific investigation or accumulation. The habit of mind good for one, is, when carried far, bad for the other. Too long study of science without applying it in practice, induces a habit of allowing knowledge to lie dormant in the mind, of regarding it as end, not as means, and to a greater or less extent, produces incapacity for applying it.

Many years ago, a foreigner was found on a work under my charge, plying the shovel and wheelbarrow, who had acquired a large amount of knowledge by years of study at a continental university. But though he knew so much, and was so expert in abstract science, he was unable to make any earthly use of it. He could not be taught to apply it to anything. In learning a giant, he was a child in everything else. This may be an extreme case, but it illustrates the tendency of all study and no practice.

On the other hand, practice keeps one on the *qui vive* to know the reasons for doing things, and the laws that operate.

Old George Stephenson, for example, picked up a great amount of knowledge, because his practice made him hungry for it, and enabled him to assimiliate it. Men of practice come to know, by what looks like intuition, things that science teaches other men only by a long course of reasoning. The habit of applying knowledge is more influential in inducing men to acquire it, than the possession of it is in inducing them to apply it.

A habit acquired in the practice of turning knowledge to account, is more valuable than a large amount of knowledge. In Franklin's time, myriads of men had much more knowledge than he had, but his habit of applying it made his little more valuable than their much.

Of course a man must have *some* science before he can apply it. But this he can get at school, or college, or in a *short* course at a technical school, while his mind is yet flexible. But a long course, reaching to a more mature period of life, fixes in the now rigid mind

a habit unfavorable to engineering success.

It can hardly be doubted that instruction in works should, when possible,
accompany that in the technical school;
just as the young lawyer or doctor learns
to practice while studying.

Too much time spent on scientific
abstractions and refinements (however
useful such things may be to the philosopher) is more than wasted by the
engineer; it unfits him for practical usefulness. Napoleon said La Place was
good for nothing for business; he was
always dealing with infinitesimal quantities.

A general ought to have been a captain
in his younger days, but if a man con-
tinues to perform captain's duty up to
the age of fifty, he is not likely to make
much of a general. So an engineer
should begin low down. But the
student should not be kept long in
acquiring mere manual skill. What he
wants is mental skill. He should be
practically familiar with iron, but it

would do him little good to be expert in making horseshoes.

It is only early practice that can teach the self-reliance, energy, and enterprise so essential to an engineer's success.

The engineer has to do with cases where the laws of nature act in different directions. Science alone cannot often give the exact resultant of those forces, sometimes unknown, often separately incapable of measurement. Experience must give the habit of estimating what allowances should be made for unknown actions and unknown quantities. Men of science once told the engineers to make fish-bellied rails, so that they should not break in the middle. The foundry laborers that broke up pig-iron could have told them that the rails would break close by the supports. Science teaches that with perfectly elastic bodies the angle of reflection is equal to the angle of incidence; practice teaches that with material bodies as they are, it never is. Time was, when for such reasons, there was some truth in the saying, that the stability of a structure was inversely as the science of the builder.

The best engineering is that which in the long run accomplishes the purpose at the least cost. The engineer should not be a mere engineer, looking only at engineering results, for then he will lose sight of their subordination to economic results. In this way so many parties have been ruined by splendid engineering. Actual practice, where money is scarce, is the best way to impress this on the young engineer. He should learn not to do, propose, or advocate anything that will not pay.

ABSTRACT OF THE ADDRESS OF PROF. FAIRMAN ROGERS.

"In my opinion the time at the disposal of the student, before he enters upon the actual practice of his profession, can be best employed in the schools, without practical work, further than the small amount which may be necessary to fix in his mind the theoretical principles which have been presented to him, provision for which can be made by very simple workshops and laboratories under the control of the professors. Beyond that, I doubt very much whether the attempt to combine practical with theoretical instruction gives an equivalent for the time spent, and I believe that the interruption of the course by a year or two years of practical work in a shop or in the field would not in the main be attended with any satisfactory result. Habits of continuous study are formed with difficulty, and should not be broken in upon until the time arrives for them to be exchanged for habits of work.

The industrious student may, with undoubted advantage, spend his vacations in each year of his study in such observations of practical matters as he may have opportunities for, a course which will result in fixing in his mind very strongly the principles presented to him by his text-books and his instructors.

There are so many things that can be taught properly only in the regular progressive methods of the schools, such as pure and applied mathematics, and mechanics, and the like, that there seems to be every reason for embracing the opportunity which can never occur again, and requiring the student to devote his time exclusively to such subjects.

Once launched into the hurry and excitement of practice, the young man finds the systematic pursuit of such knowledge difficult, if not almost impossible.

Other subjects in the same category are those based upon the digested experience of many investigators, which, though to a certain extent empirical, and wanting the logical completeness of mathematical investigations, must be adopted as embodying the principles which underlie practice.

Belonging to this class are the laws of the regimen of rivers, the action of currents, and the flow of tidal streams, or the various matters of shop or constructive practice which a man must know thoroughly at the very outset of his career, and which have been reduced to form by the labor of hundreds of individuals.

We may be assured that the young man who goes out into the world with an entirely thorough theoretical education properly given to him by competent, progressive, live instructors, will be in a position in which he cannot make serious mistakes, and from which he will surely in the long run distance those competitors who are less thoroughly prepared.

The absence of an exact knowledge of the principles which underlie practice is, I think, painfully apparent in the larger number of so-called practical men, and while we constantly hear the practical man regretting that he has not had the opportunity of obtaining that theoretical knowledge which appears to him to be so desirable, we rarely hear the man whose theory has preceded his practice complain in the opposite direction.

In a case that came under my notice

some years ago, a portion of a new building was covered with a half-span, lean-to,
iron roof, from which was suspended a
light ceiling which hid the framing from
view. With the first heavy fall of snow
of the succeeding winter the roof fell in,
and the removal of the ceiling disclosed
a curious condition of affairs, which
accounted sufficiently for the accident.

The contractor being sent for, expressed unbounded surprise, and insisted
that, as he had put up several *whole*-span
roofs, from the same drawing, of eighty
feet span, this *half*-span roof of only
forty feet ought to have been unneces-
sarily strong, and it was difficult to explain to him that, by cutting his drawing
in two, he had converted an inch and a
half round iron tension rod, which was
amply strong, into a compression piece
which was useless. In my opinion no
properly educated graduate of an engineering school, in his first year of practice, could possibly make *that* mistake,
and yet I am certain that similar things,
coming out of well-known workshops,

will present themselves to the minds of many of my hearers.

A similar case is stated to have occurred in England, where some wise individual attempted to give additional support to a whole-span iron roof which was thought to be rather light, by inserting a row of columns under the centres of the principal rafters, with the same satisfactory result.

In individual cases the precise method of education may be modified by the peculiar connections of the student giving him extraordinary facilities in certain directions, but I would sum up my remarks by saying that the best time for a young man to acquire a systematic knowledge of the fundamental principles of his science is while he is in the school, and while he is attending to what is usually called his education, and we may feel assured that he will rapidly overcome whatever temporary disadvantages he may labor under in the outset of his career for want of practical knowledge, and in a thorough and scientific manner apply those unchanging principles which have sunk into his mind and become a part of his professional nature.

REMARKS OF THE PRESIDENT, DR. R. W. RAYMOND.*

It has been a very remarkable discussion in some respects. The unanimity of feeling in one particular has been manifest, namely, as to the value of broad and general culture. This is very agreeable, because it shows that all the engineers are in favor of that thing; yet I may say that the parents in this country as a class are just the other way. When an American father talks of putting his son into any special profession he says, "I am not going to send my son to college, because he is going to be an engineer. I will take him out of college." He says, "My son is going to be a merchant; I will take him out of college;" and parents are all the time pulling their sons out of college because they are going to go into some special

* Am. Institute of Mining Engineers.

line. As I say, the tendency on the parts of fathers is exactly contrary to the tendency on the part of experts. When a man happens to be both an expert *and* a father, like my friend, Mr. Sellers, then the boy gets a wise preliminary training. But he has put his boy into his own line, and he understands what is necessary in that line. It is not difficult for a hen to bring up her chickens; it is when the hen hatches a duck that the trouble comes in; and it is the fathers who are ministers, doctors, and lawyers, who have seen some young men rise to wealth perhaps in engineering, and have got a vague notion that it would be a good thing to make engineers out of their sons, it is such fathers who are apt to think they must take off a portion of the general culture, because they fancy it does not require so much general knowledge to enter the engineering profession. They may be the soundest men on other subjects, but they know nothing about engineering.

ABSTRACT OF THE ADDRESS OF PROF. C. O. THOMPSON.

In this discussion some have held that the order should be handicraft, technics, culture; others culture, technics, handicraft; and others would arrange in other ways. But there is one objection to all these sandwiching methods. Practically we cannot hold our young men in training till twenty-five. They will go at twenty-one or twenty-two. The period of sharp acquisitiveness, the most precious part of school-life, lies between sixteen and twenty-one. Now, whichever part of a boy's triune discipline for an engineering life is allowed to usurp that period to the exclusion of the others, that will be the dominant force in his after-life. If culture, then practice will suffer; if practice, culture will suffer. Either part will be, as it were, attached to, or subordinated to, the one which " rules the favored hour." Hence it seems to me that all possible culture should be secured before a student begins his technological course, and that it should be looked to ever after. It must not be forgotten that culture is a result, or rather a growth. All we can do is to prepare the soil. The plant will assuredly grow. Perhaps, too, the best and only useful culture is to be looked for in the life for which any school training prepares a man; for I take it, we are not now speaking merely of the cultivation of the æsthetic part of man, but of that discipline of the judgment, awakening of the imagination, sharpening of perception, repression of conceit, and elevation of motive, which constitute a serviceable and efficient man of refined taste and unquestionable integrity and courage.

Let us secure as large a foundation as possible in general knowledge before the beginning of the technical course, and not lose sight of the bearing and relations of all knowledge during this course. But let us blend technics and handicraft in the technological course. The drift of this discussion has been unmistakably towards the affirmation that the technologist of the future is to be the educated workman. It is to the man whose own hands can execute, if need be, the behests of his brain, that the great engineering works of the future are to be entrusted. Engineering, so happily defined by the retiring President as "the arts of production and construction," including mechanical, civil, mining, and chemical branches, more and more condenses into mechanics. Indeed, all branches of engineering seem to react upon mechanics, forming compounds like different acids upon a common base. We are coming to think that, if a man is to be a civil engineer, he had better begin by being a mechanic. If he is to be a mining engineer, he had better begin by being a mechanic. If he is to be a chemical engineer, he had better begin by being a mechanic. This is true, at least, of all study of *applied science*.

Now, as to the amount of preliminary culture, it is desirable that at least what

is included in fitting for college should
be secured. I do not think graduates of
college in general will be drawn to technical pursuits. The whole drift of the
college is averse to them. Few boys are
so powerful polarized as the sons of the
honored member who spoke last evening.
What might be very easy for Mr. Sellers
would be very difficult for a father in other
walks. In short, it seems to be the best
available method for the average boy to
fit him for college, then send him through
a technical course in which handicraft
shall find a place; then let him enter
some manufacturing or engineering
works, and see what it all means.

ADDRESS OF MR. FREDERICK J. SLADE,
MEMBER OF THE AMERICAN INSTI-
TUTE OF MINING ENGINEERS.

MR. CHAIRMAN: If it were certain
that all the young men who enter our
technical schools had the natural qualifications necessary to be engineers, then
the problem of what the course in such
schools should consist would be materi-

ally simplified. But when we remember
that it is impossible at the early age at
which young men enter on such a course,
to determine accurately the natural bent
of their minds, the necessity of first im-
parting a liberal general education is
apparent, so that those unfit for engineering pursuits may have other fields opened
to their view, and may be drawn away
from a profession for which they have
no fitness. I therefore agree with the
remark that has already been made, that
if it were necessary to choose between a
strictly technical education and a more
general course, the latter would be the
more desirable.

I believe further, that not only is it
desirable on account of the various types
of mind to be found in a body of unde-
veloped young men, that education
should be general rather than special, in
order that none may be graduated without having received a training which
shall be of service in his particular case,
but that even were the classes composed
of none but those qualified to become

engineers, it would be much better that the instruction should be confined mainly to the theoretical part of the profession, leaving the practical details to be learned afterward, in that school of actual practice from which the engineer never graduates. I know that this would require some of our schools to give up some of the very things that they most pride themselves upon, yet I believe the effect in the end would be good.

I think it has very generally come within our experience, that those engineers who have received that very elaborate education in foreign schools, of which so much has been said, do not make the most rapid progress in the practice of their profession. The effect of their study seems to have been to give them a disproportionate confidence in the sufficiency of the formulæ and text-books of the school, to solve every problem that arises in practice, and the faculty of judgment so necessary as a check upon theoretical deductions becomes dwarfed by disuse. It sometimes seems even to be the case, that those who have received this training refuse to admit the necessity of correcting their theory by facts, and hence shut themselves out from the greatest of all schools.

It has been remarked with force by Prof. Thompson, that the acquisitive powers act with greatest vigor before the age of twenty; and he argues that on this account both theoretical and practical education should be crowded into this period. While it is no doubt true that the mind is at this age better able to receive scientific education, and to be moulded by it to correct methods of thought, it may be doubted whether the ability to weigh the value of practical expedients is as much a characteristic of a young as of a more experienced mind. It would, therefore, appear to be wiser that the period of life in which the mind is best adapted to receive that scientific training which is to shape all its future action, should be devoted to the study of science, rather than wasted in the vain attempt to acquire an incomplete and delusive acquaintance with practice. One very important result of this would be, that the young graduate could not by any possibility imagine himself an engineer, and would thus be ready to commence at the bottom, and on a true foundation lay up in intelligence and order, that experience which is the capital of the engineer.

It may be asked, why should not the teaching of practice *with* theory have the same effect in establishing a proper balance as when acquired subsequently? To which it may perhaps be answered, that the conditions under which the practical matters are presented are in the two cases widely different. In the school they are presented as problems solved, in actual practice as problems to be solved; and the fact of the solution being at hand in the former case gives a false idea of complete mastery of the profession, the precise reverse of that modesty which is forced upon one, in the latter case, by the uncertainty whether he shall be able at all to reach a solution

As to the proposition that the student should in the midst of his theoretical course, take up the study of practice as presented in the workshop, I think it is a question whether the time so spent would not be spent wastefully. It is indeed highly desirable that the young man while pursuing scientific studies, should be sufficiently familiar with at least the surface facts of practice to give life and meaning to the abstract truths which he is studying. But this acquaintance he can acquire in those afternoon visits which every young man, having a taste for engineering, is sure to make to such works as are within his reach (and in these days there are a multitude in all places), while he could acquire but little more by a constant attendance in the shop, *unless in actual employ and with a weight of responsibility resting upon him.* It is only when the sweat comes out over a man in those emergencies, when he knows that something must be done, and done quickly, that he begins to lay up valuable experience.

Now, it would be impracticable to
provide such employment for students;
and I think it may also be said that in
nine cases out of ten, where it could be
provided, the young man would never
return to school, because the ties that
would bind him to his actual work would
be too strong to be severed.

In a word, then, let the schools give a
liberal and scientific education; let the
student give concrete form to abstract
principles as he may from visits to such
works as are within reach, and by the
reading of current engineering literature,
and let the acquisition of practical knowledge begin and go on without interruption after the school course has ended.

CHAPTER V.

CONCLUSION.

The preceding pages have fully set
forth the kind of knowledge that it is
desirable should be acquired as a groundwork of engineering practice. The technical schools of different countries agree
substantially in the branches in which
proficiency must be attained by the student before the seal of their official ap-
proval can be granted to his application
to be counted as a beginner in engineering practice. In the order of importance mathematics takes the first place ;
then come physical sciences, drawing,
and language. With these some familiarity with the practical work of the profession must be acquired. So surveys,
plans, maps, and estimates form a part
of the curriculum. For the student
with ample means the way is clear. He
may select his school and follow the prescribed course. Having completed it, if

he is content to accept a humble situation and will patiently wait for promotion until his experience shall have rendered his services valuable, his introduction to the higher duties and responsibilities of engineering life is assured, unless indeed he has entirely mistaken his profession.

For him who must earn his living before he is able to complete such a course of study as is universally pronounced requisite, the way to proceed is not so clear, especially if the occupation he is obliged to follow is not of a kind to bo in itself of service as a means of education in the line of his chosen profession. As in such a case he must depend upon his power to acquire knowledge in his leisure hours, and complete by himself the studies which his more fortunate competitor is aided to accomplish. If he has completed a fair academic or high school course before he enters upon daily service, as an earner of wages, his condition is far from unfavorable, especially if he has fully determined to overcome obstacles in the way of acquiring knowledge. If the daily occupation be that of an artisan, it is an open question whether his training is not of the best kind to insure success. For the education he needs is of two kinds—theoretical and practical ; and the period of life most favorable to the acquisition of either kind is between sixteen and twenty years of age. It is not of so much importance that the trade at which he works should be directly related to engineering practice. The work of any artisan methodically and steadily pursued under skilful guidance, is an education of the eye and hand and judgment of the highest value. The opinions of men eminent in the profession might be referred to, which are quoted in the preceding pages, which are quite pointedly to the effect that the best order of preparation is first the school, then the shop, and then the finishing studies. The reasons given for this order are that the training of the hand and the eye, as well as that of the mental powers, should be brought within the period of easy acquisition ; also that the eagerness to learn through study is much stimulated by alternation of school and shop, and so the accomplishment of the final studies is rendered easier.

Observation of the pupils in the night-schools of our great cities certainly justifies the belief that daily labor in the shop in no sense unfits the learner who has once been a student, for steady progress in acquiring knowledge. Prominent engineers now living have achieved their successes without other aid from instructors than that obtained in the public day-schools and the night-school. The higher technical schools afford the best means of entering upon the professional work of the engineer ; but other routes are open, and with patience and persistence the goal may be reached.

It should be borne in mind that the learning period ends only with life. The graduate of the Engineering School can rightfully regard his degree only as a certificate that he is prepared to become an engineer.

One part of the education of the young learner has not, it seems to the writer, been sufficiently urged as a necessary part of the school training. That is, some branch of natural science ; one which will stimulate a search for specimens : either entomology, botany, or mineralogy (or lithology). The latter is probably to be regarded as the best when the direct bearing upon the after practical experience is considered. Nothing in the way of education so trains the powers of observation as the search over field and through forest for *specimens*. No phenomenon escapes the notice of the young naturalist. If he be a collector of minerals or rocks, every ledge and every bowlder is an object of interest, and his pursuit takes no account of seasons. It is of inestimable advantage to the engineer that the minute features of a structure or a landscape are seen at a glance ; but such seeing depends upon early training and practice under the stimulus of the pleasurable excitement which always accompanies out-door searching for specimens.

It is a natural sequence to this course of training that the learner becomes a student of natural phenomena, and the action of streams, the decay of rocks, and the changes wrought by tides and winds are to him subjects of absorbing interest. Every rivulet is to him the analogue of a great river, and will exhibit, either spontaneously or through slight artificial constraint, all the features of erosion, transportation, and deposit, forming bars, natural levees, and deltas which are only miniatures of those greater ones which at times become the anxious consideration of the experienced engineer. The rock whose surface crumbles under the exposure to rain and frost, be it ever so little, is noted and regarded as worthless to the builder. The dunes formed even though temporarily by the wind along sandy shores are studied as likely to afford valuable hints in regard to regulating and controlling those larger accumulations of similar kind which sometimes prove troublesome to the peace and comfort of a community.

It would be easy to continue this line of advice to much greater extent, but the proposed limits of this little treatise forbid. For hints about pursuing this kind of out-door observation, an excellent aid is to be found in a small work by the eminent author De La Beche, entitled "How to Observe."

It can hardly be necessary to urge the student who designs to be an engineer to carefully observe the progress of structures in course of construction, and especially to note the little devices by which weights are raised, structures are stayed, streams deflected, etc. The object of the last few pages has been to advise such a training of the perceptive faculties that the smallest of the expedients of an engineering work could not well escape notice.

That the broadest possible culture is at least desirable in an engineer has been aptly urged by more than one of the eminent speakers whose expressed opinions are quoted in the foregoing pages. That he should cultiv.ite a knowledge of business habits as early as possible is sufficiently evident from the fact that his dealings are largely with the leading business-men of the community. The same consideration suggests that a reputation for the strictest integrity is as essential to him as to the banker or the judge, and for like practical reasons.

APPENDIX.

Extracts from an address before the American Society of Civil Engineers at its First Annual Convention, in 1869, by John B. Jervis, Honorary Member of the Society.

All professions must have a beginning. There must be preparation, just as when a fabric is to be reared, the rubbish must first be removed and a proper foundation prepared on which the structure is to rest. You will suspect, from this, I am going to talk to the beginner, rather than to those who have a matured skill and experience.

A civil engineer I understand to be a man who devises and executes works, as canals, railroads, water-works, bridges, mining, etc. I regard it as different from the strictly mechanical engineer; but the civil engineer should know much of mechanical engineering, though he be not a professor in devising and constructing machinery. ·

After a fair education in the ordinary elements, the young man that designs to prepare for the profession of an engineer should study mathematics so far as to qualify him to make any computation of quantities, and to carry forward any investigations that he may find it necessary to make in pursuing the science of mechanical philosophy, and especially in regard to the strength of all materials that may be required in the structures he may be called upon to erect, and the capacity of his structures to support the object for which they may be designed.

The engineer, having got thus far in his study, he is prepared to enter on the study of mechanical philosophy. In this branch he will hardly be able to learn too much. It enters deeply into the affairs of an engineer. In this, especial attention should be paid to the character of all the materials required in the varied structures it is his business to provide for, and the form and position of materials best adapted to the end it is sought to secure.

The next object I should propose for study is hydraulics. This enters into nearly all the questions in engineering, especially canals, water-works, mining, and bridges. In some respects it is a difficult science, not in its mathematic or theoretic aspect, but in the difficulty of obtaining accurate data on which to rest the computations for specific objects. This difficulty only renders its careful study more important, and it will be found that much has been reduced to a scientific form, and if not in all cases exact, is so close an approximation as to afford a reasonably safe guide in practical hydraulics.

I have not noticed the surveying feature, for, with a proper knowledge of mathematics, it is only necessary to learn the use of instruments in order to establish lines and levels. The education, thus far, may be obtained at any school that has a good mathematical department.

The next step in education should be the study of structures of various kinds that have been erected by experienced engineers. These have been published in various books, and may be advantageously studied without a professed teacher. So far, the student is merely a student, and is only well prepared to enter on the practical or real study of his profession.

Every work detailed and set forth in books has been erected with some specific object, and under circumstances that are not likely to be in all respects found again for any similar work. The young engineer will, therefore, require great care in recommending a structure he has examined in a book, to see that the circumstances surrounding his case are analogous to the original. Neglect, in this respect, may lead to very unfavorable results. With a sound knowledge of principles in regard to form of structure best calculated to secure the object for which the work is designed, and to afford the necessary stability and permanence, and of the materials best adapted to this purpose, he will be able to make an intelligent criticism on what he examines, and judge how far it may be a guide in his operations.

The next step in this training is to enter on the field of practical duties under the guide of an experienced engineer. In this situation he will have opportunity to examine the unbroken ground, learn the reasons for the various parts of the work, and why it may be proper to depart from the order of some similar work he may have noticed in his studies. Here he will see the ground to be occupied, and, by communion with his principal, will see how far it may be adapted to the proposed work—the various needs of the structure, the facilities for materials most suitable, and how far a modification in plan and materials may be required to meet the circumstances that may exist.

No skill in forming lines and levels, and in devising structures, will complete the education of an engineer without an intelligent capacity for conducting business. This is an important item in his education, and indispensable to a successful practice. The training of his mind should make this an easy acquisition. He will often find it necessary for his own investigations to make written statements and tabulations, which will call into action his skill in arrangement —even his common note-book will exercise his powers to put his work in an intelligible and convenient form. If he keep in mind that he should be prepared to protect himself against forgetting his own work, and be able, at any time, to give an account of his doings, he will feel the necessity of order and ready reference, which are the essential elements of business. Some people suppose an engineer, as a matter of course, knows nothing about business management. An absurd mistake. No profession more needs thorough business qualification. It is still a further error to suppose an engineer, by his education, is unfitted for systematic business. The fact is the very reverse. His education peculiarly qualifies him for a systematic management of business, and his professional duties demand the most varied knowledge in this respect. The first operation in his business life is to set forth, in intelligent form, the work he has been doing; a line, a level, or a series of computations to be set forth in order. Then he comes to prepare contracts and specifications for work, which demands an accurate knowledge of all he wants done, set forth in items so described that there will be no misunderstanding; an important business matter. His work now under contract, he must measure and compute all items of work, and these must be so arranged, in a suitable book, that they can be referred to, and made the basis of his occasional and final statements of work done. So far I have only treated of that branch of a business education that relates to the method of preparing tables, accounts, and statements. . . . The most important feature in business is still to be considered, namely, a knowledge of men and of the value of work. I know of no occupation that better qualifies a man in this respect than engineering. He has to deal with men who, as a class, are proverbially sharp in the conduct of their affairs, with whom many questions arise that are not to be determined by a simple computation, and even computations will be questioned. With these men the engineer holds a delicate relation, as the umpire between the contracting parties; and will often be placed between plausible claims on one hand, and a sense of duty on the other. In these circumstances he will have great opportunity to obtain a business knowledge of men. Some he will find upright, though they may have mistaken views of their rights; and others he will find under much pretension to seek what they should not have. Under various conflicts the engineer must aim to do justice between the parties. They have committed to him the duty of adjusting all questions, and in this he must examine the bearing of all claims, and though he may be annoyed at what he thinks an unjust demand, he is in duty bound to render equity, according to the terms of the contract. The engineer being in the employ of one of the parties, it is indispensable that he maintain the reputation of an upright man, for on this the contractor must depend. If he shows a disposition to take undue advantage, not warranted by the terms and spirit of the contract, the contractor will lose confidence if this is against him, and, if in his favor, the other party will be dissatisfied. In all such cases, committed to the judgment of the engineer, he will need the best experience as a business man, and especially to cultivate the golden rule of doing as he would be done by. It is always prudent for an engineer to so prepare his specifications of work that no misunderstanding may arise. But this cannot always be done, as contingent work is sometimes, and I may say often, required. . . .

The business capacity of the engineer will, in most cases, be tried in his intercourse with the managers of the enterprise. In the utmost good feeling they will be very apprehensive of expense, and desire various methods of reduction, which will appear plausible to them, as well adapted to their circumstances. The engineer, very naturally desires to make a work that will be creditable to his profession. But he must listen respectfully to the suggestions of his principals, and, so far as he can, modify his plans to meet them. The engineer may be well satisfied his plans are best adapted to the permanent interest of the enterprise, and, so far as he is able, should convince the managers they are so. In this he will find need of all his business tact, to yield when circumstances, and especially the want of funds, afford a reasonable apology, or even a necessity, for so doing. It may not, in fact, be practicable to make his work as permanent or complete as its true interest requires, for want of funds, or for want of the most suitable materials. Of these he must exercise his discretion as a business man, as well as a skilful engineer.

There is another point in this connection that will often be more trying than the above. The managers of such enterprises have been known to conduct their affairs with a view to make them subservient to their private interest or ambition. They may do this more or less, without interference with the duties of the engineer, in which case he may know nothing of it, or, if does know, he may not be bound as a duty to take notice of it. But it is very likely to come in conflict, and he will be expected to shape his professional duties in a way that will promote private interests at the expense of the institution. Great skill and adroitness will be practised, and if the engineer has any weak side it will surely be found.

These things will always be done under profession of serving the institution. To avoid wrong-doing, and, at the same time, give no just offence under the circumstances, will surely try the business capacity of an engineer. The matter in issue may bring a crisis that will compel resignation.

In review of the brief discussion I have given, I think I am fully warranted in the opinion that the training and practice of an engineer should make him peculiarly eminent as a business man, not less than skilled in designing and erecting works in his profession.

In railway building, for the most part, the engineer has been regarded merely as an expert, to run lines and levels, and compute quantities. When the work, in this respect, was done, he was regarded as of no more use, and to retain one of his assistants to do a little levelling or surveying, for some contingent work, was all that was regarded as necessary from his department. No doubt some engineers who have built railways had

no professional qualification to go further. No one can complain of them;
they did all they supposed required of
their profession, and this was, perhaps,
all their employers expected of them as
engineers. It may be said why not employ an engineer of more experience; to
which it may be replied, they probably
were not to be had. It has been a great
error on the part of engineers and pro-
prietors that the impression to a great
extent prevailed that in regard to machinery, station grounds, buildings, and
shops, the engineer gave way to the
mechanic. This circumstance has put
back the profession, keeping the engineer
from the proper study of these duties,
so that only a small comparative number
have given proper attention to this important branch of professional duty. So
long as the engineer did not assume
these duties, the resort to the mechanic
was very natural, and he would be regarded as best qualified to make up the
complement of the railway. The excuse
for this is, the profession in general had

little knowledge of what was wanted in these respects, and the engineer was regarded as a sort of refined surveyor, and as knowing nothing about engines, cars, and shops. Now, the professional point is, that the engineer's education should qualify him better than any other to provide for everything about a railway.

The engineer constructs a railway, and should know better than any other what sort of machinery is best adapted to its use. Of course he will go to the mechanic for his engines, cars, etc., but he should know the general character, and be able to specify the leading or principal characteristics of what he wants. Certain accommodations will be wanted for the current use and repair of the machinery, which requires much thought and care on the part of the engineer to secure the greatest convenience and economy in the operations of business. In most cases it will be necessary to keep in view the enlargement of stations and shops, to provide for a probable increase in traffic, and, while building for present wants, take care to provide, by method, for enlargement at a future day, with grounds secured for this purpose, so that no pulling down shall be required. In the history of our railways, it is palpable these things have been imperfectly considered. It has been stated why our works have been subject to these imperfections ; but that does not excuse their want of improvement. One reason for neglect may be found in the fact that, for the most part, engineers did not regard this as properly belonging to their profession, and when a railway was put in operation they withdrew, as having accomplished their mission.

The engineer eminently depends on character. The interests of others, in various ways, are committed to him. On his capacity for his profession, and his integrity as a man, reliance must be placed. He will meet many difficulties of a physical, and not a few of a moral nature. He is in a progressive calling, and has occasion to be constantly learning. I suggest a ready observation of what he may see, a constant study and reflection on the varied duties of his profession, and a watchful guard against committing himself until he is fully prepared to set forth his views clearly and decidedly. To neglect the latter will be very likely to embarrass, and may defeat the object of his labors.

Suggestions Offered to the Council of the Institution of Civil Engineers on the Subject of Engineering Education.

FROM SIR JOHN RENNIE, F.R.S.

Any person intended for the profession of a civil engineer, in my opinion, should be educated as follows: In the first place, he should be sent to some good school until about twelve years of age, where he may be thoroughly grounded in reading, writing, composition, arithmetic, algebra, geometry. English, Latin, and Greek grammar. I mention the two latter, as it will enable him hereafter to acquire with facility European or other languages as occasion may require, particularly French, German, Italian, Spanish, etc.

From about twelve to sixteen he should be sent to one of the junior universities, such as King's College, the London, Edinburgh, Glasgow, or Dublin Universities: here he should attend the different classes of mathematics, algebra, geometry, plane and spherical, trigonometry, astronomy, natural philosophy, geology, geography, chemistry, electricity, and drawing. assisted by private tutors, who, by frequent examinations, should make him thoroughly understand what he has learned, and enable him to pass with credit the several public examinations which he must undergo.

At the age of sixteen, if he has been *diligent* and *well looked* after, he ought to have obtained a sound general education, particularly as regards the scientific department connected with the profession of a civil engineer.

At sixteen he should be apprenticed to some practical manufacturing engineer of eminence, where he should commence working with his hands, and go through all the departments of pattern-making, founding, turning, fitting and erecting steam-engines, marine, locomotive, and fixed, and all the variety of machinery connected with them and railways, and a general knowledge of ship-building, whether of iron or wood, also the mechanical drawing and calculating departments connected with them.

After having been well employed in this manner for three or four years, combined with his previous education, he ought to be well versed or grounded at least in mechanical engineering, which forms one of the most important parts of the education of a civil engineer.

Having completed this he should be sent to some good scientific and practical nautical and land surveyor for a short time, where, if diligent, he will soon be able to acquire a knowledge of levelling, laying out lines of roads, railways, canals, drainage, and mapping large districts of country and sea-coasts.

He should then study practical hydraulics upon a great scale, such as the principles and management of rivers, embanking, draining, sewage, water supply, irrigation, the planning and construction of harbors, docks, bridges, light-houses, masonry, carpentry, earthwork, etc., with a thorough knowledge of the use and best mode of applying materials of all kinds; and having previously obtained a knowledge of working iron and timber, he should do the same with stone, cements, and other materials. He should likewise practise himself in drawing up detached reports, plans, specifications, and estimates of the various works. In these departments he will be greatly assisted by reading and carefully studying the reports and plans of Smeaton, Telford, Stephenson, Brunel, Watt, Rennie, and other eminent engineers.

He should also study architecture, so as to be able to design and construct all the buildings connected with civil engineering, such as railway stations, sheds, and warehouses.

Public buildings and ornamental architecture, strictly speaking, are rather out of his line ; nevertheless, if he has time and taste for it, he can do so at his leisure.

With regard to languages, it certainly is desirable that an engineer should know well German, French, and Italian; having previously laid the foundation of them, he can easily acquire them at his leisure.

With regard to the higher classes of physics and mathematics, these, although not absolutely necessary for the practice of civil engineering, still, no doubt, they form a most valuable accomplishment for those who have time and taste to acquire them. With regard to the honors conferred by purely scientific societies, they are open to any one who chooses to qualify himself for them, and desires them; and if either his executed works or writings are worthy of these distinctions, he will have no difficulty in obtaining them.

After a candidate for civil engineering has creditably gone through the above education, he will have arrived at about the age of twenty-three or twenty-four years; he may then confidently present himself to any engineer in large practice as an assistant at an adequate salary, or he may practise upon his own account; but he must always bear steadily in mind, *that nothing but constant industry and hard work, combined with a thorough determination to overcome all obstacles* in his progress, can make him a good engineer. I should not recommend Eton, Harrow, Oxford, or Cambridge; they may do for the law, medicine, or the Church, not for the engineer.

The above course of education appears to me the best adapted for a civil engineer, and if the candidate will only pursue it zealously and steadily he may in time become a credit to the profession.

FROM MR. J. M. HEPPEL.

What seems to lie at the root of the question is to define the character and extent of knowledge which it is desirable a young engineer should possess previous to entering on the regular and responsible practice of his profession. On such a subject no doubt much diversity of opinion will be found to prevail, as well as on the best modes of acquiring such knowledge as is by general consent admitted to be requisite. I take it that I shall best meet the wishes of the committee by a simple statement of my own view.

To begin with, I think it extremely desirable that the young engineer should possess all the ordinary acquirements of a well-educated gentleman. He should be a moderately fair classical scholar, and should understand French and German, and if also Italian and Spanish, so much the better. In English, besides a fair acquaintance with history and general literature, he should be master of composition so far as to express himself in speaking or writing with precision and force. Coming now to what may be considered more special, he should have a sound knowledge of common geometry, so much of trigonometry and conic sections as is usefully applicable in practice, of arithmetic, the nature and use of logarithms, of ordinary algebra, and of the higher analysis, so far as the general principles and most useful applications of the differential and integral calculus, a complete acquaintance with the theory of statics, and so much at least of dynamics as will enable him fully to understand any ordinary treatise on the action of the chief motive powers, as gravity, water, or the various forms of heat. He should be fairly acquainted with chemistry, geology, physical geography, and most of those branches of science which are commonly grouped together under the name of natural philosophy.

Even yet we have come to nothing which may be said to belong to the special training of an engineer, being rather acquirements which would be desirable and useful to most persons, and which are more or less possessed both by private gentlemen and members of other professions: those which follow have more special application.

The young engineer should have a good practical acquaintance with the mechanical properties of the principal materials of construction, and a sufficient knowledge of the trades of a carpenter, smith, millwright, bricklayer, and mason, to be a competent judge of their products. He should be a good mechanical draughtsman, and able to sketch and design clearly by hand.

He should have a thorough practical acquaintance with surveying, with the setting out of works from plans and levelling, and with the construction and adjustment of all instruments used in these operations. He should be expert in taking out quantities from plans and framing estimates, and be a sufficiently good accountant to fully understand and control accounts of expenditure in works.

Lastly, he should, both by reading and observation, and, if possible, by actual charge and responsibility, have collected a good body of knowledge of the best practice at home and abroad in executing important and difficult works.

The list appears long, but I doubt if it is by any means complete. I think, however, with these attainments, or a good proportion of them, we should have the making of a very fair engineer, and this is all that can be expected from education; the rest must follow from the genius, perseverance, opportunities, and good fortune of the individual.

Looking now back to my list, I think it will be apparent that all the literary, mathematical, and purely theoretical matter, including some knowledge of chemistry, geology, geography, and natural philosophy, would be easily acquired by a boy of fair abilities at a public school in five years (say from ten or eleven to fifteen or sixteen), followed by one or two years' attendance on lectures at University or King's Colleges. At least so much would be acquired that the rest might easily be made up by private study concurrently with the subsequent pursuit of the more practical part.

Mechanical drawing should also, if possible, have been to a certain extent acquired during this period; but if not, a short time should be given to it specially, including the making of accurate and detailed sketches, with dimensions of machinery and works, and then drawing them to scale. As soon as the pupil has attained to some degree of proficiency in this, he should give from one to two years to the learning of some mechanical trade, amongst which I should prefer those of a smith or millwright: he should work full hours with the men, and as soon as possible earn wages.

The order in which the rest of the subjects I have mentioned are acquired does not strike me as very important, and the pupil might profit by the best opportunities. They might, as it seems to me, all be learnt in the office of an engineer, always supposing two things:

first, that his practice is sufficiently extensive; next, that the efficient education of the pupil, and not the mere utilization of him, is steadily kept in view. No doubt there are objections. The office of an engineer is not, and probably cannot be made, an educational establishment. In the first place, he has too much personal occupation to be able to attend much to pupils; and in the next place, where work has to be got through, a pupil will inevitably be more or less kept to what is most pressing, or what he can do best. But supposing the most conscientious regard to be paid to his interests, even with the most extensive practice, opportunities must be waited for, whilst in a more restricted one they may be entirely wanting.

It is chiefly this latter class of considerations which lead me to insist on so much preliminary knowledge before entering an engineer's office. A young man thus fortified would be very differently circumstanced from a boy just taken from school, and would probably find few things intrusted to him, however unpromising, from which information and improvement could not be derived. He would himself understand his own wants, and would not allow opportunities to be lost simply from inability to understand and appreciate them. Some might be more fortunate than others in the extent and variety of subjects presented to them; but I think, with moderately fair treatment, very few could fail in the course of three or four years to obtain such a body of information as would enable them to evince their capacity for responsible charge of works.

Leaving now for a moment this plan, which is the one usually followed more or less in this country, let us consider another, partially adopted here, but much more extensively on the Continent—that of colleges or complete establishments for the training of civil engineers. As far as the purely scientific and literary part of the course goes, there seems to be no reason why such a college should not impart as sound and complete instruction as any similar institution, supposing always the basis of a single profession to be sufficient to maintain it on a satisfactory scale; but then for this alone a separate establishment seems hardly requisite, as there appears to be nothing but what can be supplied by existing schools and colleges, and supplied, as it seems to me, if the courses are properly selected, so as to give full occupation to the student in the acquisition of what he really requires without superaddition of redundant or superfluous matter. If we look, however, beyond this to the more special training, I think some marked disadvantages, as compared with the former plan, will become apparent.

The work is not real work. The plan of the bridge may be neatly and carefully drawn, its strength calculated, its cost estimated; and all this may be submitted to judicious criticism, but nobody has got to build it. A survey may be made of the neighborhood of the institution, or of a line of road or farm in its vicinity, and nicely plotted, but it has not got to pass a Committee on Standing Orders. A line of canal may be laid out and levelled, and a working section and detailed plans prepared, but it is exempt from verification by that most uncompromising examiner—water.

I do not at all mean to deny that even in these practical branches much useful information might be acquired, and the advantage of being able to instruct in classes and of selecting the order of subjects is obvious; but I do think that this kind of instruction cannot obviate the necessity of a subsequent probation in a real office and in actual work; and in many cases it might possibly have been there acquired as rapidly and certainly more completely.

In my point of view, then, the question is brought within a very narrow compass. The engineer's education should begin at a good school, best at a public one. Following this should come a college course, more or less prolonged. He should then master some mechanical trade, or the order of the two last might, if more convenient, be inverted, and should, in any case, finish with a certain term of pupilage under a practising engineer.

EXTRACT FROM SCOTT RUSSELL'S WORK ON TECHNICAL EDUCATION.*

What Technical Education should we give to the Mechanical Engineer or Machinist?—From the days of James Watt and Arkwright until now, comprehending the whole of the present century, the mechanical engineer or machinist has formed one of the most important classes of this country, and has conferred on it immeasurable benefit. It was the mechanical engineer and the manufacturer who, together, during the early part of the present century, while the whole of Europe was overrun by the

---

* Systematic Technical Education for the English People. By John Scott Russell, F.R.S., etc. London, 1869.

curse of war, created wealth in this country so rapidly as to enable her to struggle through a burden of expenditure to which there has been no parallel, and to come out of it prosperous and wealthy.

There are no occupations or trades concerning which there could be so little difference of opinion as to the practical importance of special technical education, as this class of mechanical engineer and machinist. Philosophers have defined man as the tool-using animal; but if the man of this century were defined, the "engine-maker" and "machine-user" would be his leading characteristic : it is the triumph of human nature in our time, that it has achieved the understanding of the forces of nature so completely, that whatever material service we wish to perform, we can always discover some elementary force in nature willing to lend us its aid to conquer our difficulty, prouided we will study its nature sufficiently to direct it into the way in which it can best serve

our end. The steam-hammer of Nas-
myth, and the steel ingots of Krupp, are
symbols of the powerful yet plastic forces
man wields, in his gigantic shape-com-
pelling processes of manufacture. We
may sum up the duties of a man of this
craft by saying that there is scarcely a
process now performed by animal or man
which our engineers or machinists of the
next generation may not be called upon
to perform better and quicker by machines of their own creation.

Of the engineer and machinist it is
therefore very easy to indicate the course
of instruction ; unluckily, much easier
to indicate than to accomplish. He must
master all the known powers of material
nature: heat and cold ; weight and im-
pulse ; matter in all conditions—liquid,
solid, and gaseous ; standing or running, condensed or rare, adamantine or
plastic—all must be seen through and
comprehended by the master of modern
mechanics. The same laws which gov-
ern the machinery of the heavens, he
has to apply to the machinery of the

earth ; and the same exquisite mechanism which the Creator has used in the structure of his animals, the modern mechanician has to apply in the construction of his microcosms. The modern mechanician who would be equal to his work must be prepared to shape a tool and frame an engine for the execution of tasks which were never even dreamt of by the older mechanicians.

Technical Education of the Mechanical Engineer and Machinist

Mechanical Knowledge.	*Technical Education.*
Shapes and sizes of things	Geometry.
Quantities	Algebra.
Numbers	Arithmetic, Calculation.
Weights	Laws of Gravity.
Forces and Motions	Laws of Dynamics.
Strengths	Laws of statics.
Mechanical powers	Theoretical mechanics.
Laws of solids	Kinematics.
Laws of liquids, Laws of airs and gases,	Hydrostatics and hydrodynamics.

Mechanical Knowledge.	Technical Education.
Generation of motion .	Heat, light, electricity, attraction, and repulsion.
Sources of power . .	Chemical physics.
Applications of power .	Elements of mechanism.
Mechanical inventions .	History of machinery.

What Technical Education should we give the Civil Engineer?—The great public works of a civilized country have always demanded and generally received from its Government earnest solicitude and forethought. In France, the civil engineers are the *élite* of the nation; the most distinguished pupils in the colleges throughout the country are promoted into the central technic institution of France in Paris; and out of this again, a selection is made of the most talented for the "corps de génie maritime;" for the "corps de génie militaire;" and for the "corps de génie civil," or "ponts et chaussées."

By the great public works of a country so much is gained or lost to the public

well-being, that the most liberal measures are justified if they succeed in pro
viding for its service the profoundest
knowledge, the most brilliant talent, and
the highest skill.   In the time of the
Romans, Europe was covered with those
wonderful roads which have been per-
petuated to the present day, and are
marvels of conception and execution.
The correction of rivers and supply of
waters to great cities, the drainage of
marshes and the irrigation of plains,
have developed the industry and created
the wealth of populous countries; and
it has depended almost entirely on the
wisdom or folly of modern Governments,
in the selection of their engineering systems, whether those great engines of
commerce, the modern railways, have
been given to a country at small cost, on
a wise system of development, with gain
at once to the capitalist, to the trader,
and the Government.   Where Governments have been wise, the railways have
been well selected, cheaply made, eco-
nomically and profitably worked.   Where

they have been reckless, ignorant, unwise, railways have been made at great cost, extravagantly worked, dear to the public, and unprofitable to the capitalist.

When it is considered that the telegraphs which now work the commerce of the world ; the great lines of steamships which unite its most civilized portions ; the railways which everywhere connect the populous centres of empires ; the water supply ; roads ; ports and harbors; the direction, training, and permanence of our navigable rivers—are all works involving enormous cost, involving the highest national interests, and requiring consummate knowledge and skill, it is plain that we may judge of the wisdom of a nation by the foresight and forethought it bestows upon the rearing, training, and selection of this *corps d'élite* or *corps de génie ;* and it is therefore self-evident that, in a technical university, the pupils of this section must find a prominent place. For England especially, with her wide-spread dominious, it is evident that the youthful engineer should be prepared to find a sphere of usefulness in any quarter of the globe, and to carry with him a mastery of all the resources of modern science and skill.

Engineering Knowledge.	Technical Education.
Laws of water—standing and running.	Hydrostatics and hydraulics.
Laws of dead matter . .	Strength and resistance of materials.
Laws of fixed and moving bodies	Statics and dynamics.
Building of bridges and ways.	Theory of structures in stone, timber, and iron.
Surveying, mapping, selection of routes.	Geometry, trigonometry, and surveying.
Erection of buildings . .	Theory of beauty and ugliness.
Estimates of cost and production of public works.	Prices, wages, and economical valuations.
Steam-ships and machinery.	Naval architecture and mechanical engineering.

To a great extent, the civil engineer must have also the same education as the mechanical engineer.

The proposed studies are thus defined :

THE SCHOOL OF MECHANICS.

Pure Science.

Higher Geometry.	Higher Dynamics.
" Algebra.	" Energetics.
" Arithmetic.	" Chemistry.
" Statics.	" Metallurgy.

Practical Applications.

Descriptive Geometry.
Constructive Geometry.
Geometric Movements.
Sources of Materials.
Properties of Materials.
Strength of Materials.
Elements of Mechanics.
Structural Mechanics.
Machinery and Tools.
Engines and Prime Movers.
Economics of Work.
Endurance of Machinery.
Machine Shops and Buildings.
Mechanical Manufactures.
Political Economy.
Workshop Economy.
Principles of Design.

Work.

In the Drawing Office.
In the Collection of Machines.
In the Collection of Machine Materials.
In the Collection of Raw Materials of Manu-
factures.
In the Collection of Engines, etc.
In Mechanical Experiment.
In the Factory.
Round the Tour of Home Manufactories.
In Foreign Travel.

THE SCHOOL OF CIVIL CONSTRUCTION

ENGINEERING.

Pure Science.

Higher Geometry.	Higher Energetics.
" Algebra.	" Hydrology.
" Arithmetic.	" Chemistry.
" Statics.	" Geology.
" Dynamics.	" Crystallogy.

Practical Applications.

Engines and Prime Movers.
Theory of Vehicles and Locomotive Ma-
chines.
Theory of Ships and Steam-boats.

Chemistry of Building Materials.
Geology of Stones and Cements.
Mineralogy and Metallurgy.
Stability of Foundations.
Building Combinations of Materials.
Sources of Materials of Construction.
Theory of Bridges, Roofs, and Tunnels.
Constructive Geometry.
Graphic Geometry and Surveying.
Descriptive Geometry.
Perspective Geometry.
Geometric Movements.
Strengths of Materials.
Elements of Mechanics.
Machines and Tools.
Theory of Rivers.
Theory of Tides and Waves.
Theory of Roads, Railroads, and Canals.
Principles of Architectural Design.
Principles of Metallurgy.
Economics of Construction.
Endurance of Structures, Engines, Machines, and Implements.

Work.

In the Drawing Office.
In the Collection of Engineering Models.
In the Collection of Building Materials.
In the Collection of Machines
In the Laboratory of Strength of Materials.

In the Chemical Laboratory.
In Engineering Experiment.
In the Factory.
On the Works.
In Foreign Travel,

HERRMANN, GUSTAV. The Graphical Statics of Mechanism. A Guide for the Use of Machinists, Architects, and Engineers ; and also a Text-Book for Technical Schools. Translated and annotated by A. P. Smith, M.E. 12mo, cloth, 7 folding plates...........$2.00

HENRICI. OLAUS. Skeleton Structures, Applied to the Building of Steel and Iron Bridges. Illustrated........................$1.50

HOWARD, C. R. Earthwork Mensuration on the Basis of the Prismoidal Formulæ. Containing Simple and Labor-saving Method of obtaining Prismoidal Contents directly from End Areas. Illustrated by Examples and accompanied by Plain Rules for Practical Uses. Illustrated. 8vo, cloth...............$1.50

JOYNSON, F. H. The Metals Used in Construction. Iron, Steel, Bessemer Metal, etc. Illustrated. 12mo, cloth............. ...... .75

——Designing and Construction of Machine Gearing. Illustrated. 8vo, cloth $2.00

KANSAS CITY BRIDGE, THE. With an Account of the Regimen of the Missouri River and a Description of the Methods Used for Founding in that River. By O. Chanute, Chief Engineer, and George Morrison. Assistant Engineer. Illustrated with 5 lithographic views and 12 plates of plans. 4to, cloth, $6.00

MERRILL, Col. WM. E., U.S.A. Iron Truss Bridges for Railroads. The method of calculating strains in Trusses, with a careful comparison of the most prominent Trusses in reference to economy in combinations, etc. Illustrated. 4to, cloth. Fourth edition..$5.00

MORRIS, E. Easy Rules for the Measurement of Earthworks by means of the Prismoidal Formula. 8vo, cloth, illustrated......$1.50

PLYMPTON, GEO. W. The Aneroid Barometer : its Construction and Use. Compiled from several sources. 16mo, boards, illus....... .50

POCKET LOGARITHMS, to Four Places of Decimals, including Logarithms of Numbers, and Logarithmic Sines and Tangents to Single Minutes. To which is added a Table of Natural Sines, Tangents, and Co-Tangents. 16mo, boards........................ .50

RANKINE. W. J. MACQUORN, C.E., LL.D., F.R.S. Civil Engineering. Comprising Engineering Surveys, Earthwork, Foundations, Masonry, Carpentry, Metal-Work, Roads, Railways, Canals, Rivers, Water-Works, Harbors, etc. With numerous tables and illustrations. Seventeenth edition. Crown 8vo, cloth.................................$6.50

——Machinery and Millwork. Comprising the Geometry, Motions, Work, Strength, Construction, and Objects of Machines, etc. Illustrated with nearly 300 woodcuts. Sixth edition. Crown 8vo, cloth...........$5.00

——Useful Rules and Tables for Engineers and Others. With appendix, tables, tests, and formulæ for the use of Electrical Engineers. Comprising Submarine Electrical Engineering, Electric Lighting, and Transmission of Power. By Andrew Jamieson, C.E., F.R.S.E. Seventh edition. Crown 8vo, cloth....$4.00

——A Mechanical Text-Book. By Prof. Macquorn Rankine and E. F. Bamber, C.E. With numerous illustrations. Third edition..$3.50

RIPPER. WILLIAM. A Course of Instruction in Machine Drawing and Design for Technical Schools and Engineer Students. With 52 plates and numerous explanatory engravings. Folio, cloth.......................... ..$7.50

ROEBLING, J. A. Long and Short Span Railway Bridges. Illustrated with large copperplate engravings of plans and views. Imperial folio, cloth........................$25.00

SCRIBNER. J. M. Engineers' and Mechanics' Companion. Comprising U. S. Weights and Measures, Mensuration of Superfices and Solids, Tables of Squares and Cubes, Square and Cube Roots, Circumference and Areas of Circles, the Mechanical Powers, Centres of Gravity, Gravitation of Bodies, Pendulums, Specific Gravity of Bodies, Strength, Weight, and Crush of Materials, Water-Wheels, Hydrostatics, Hydraulics, Statics. Centres of Percussion and Gyration, Friction Heat, Tables of the Weight of Metals, Scantling, etc., Steam and the Steam-Engine. Nineteenth edition, revised. 16mo, full morocco...$1.50

SHIELDS, J. E. Notes on Engineering Construction. Embracing discussions of the principles involved, and Descriptions of the Material employed in Tunnelling, Bridging, Canal and Road Building, etc. 12mo, cloth..$1.50

SHREVE, S. H. A Treatise on the Strength of Bridges and Roofs. Comprising the Determination of Algebraic formulas for Strains in Horizontal, Inclined, or Rafter, Triangular, Bowstring, Lenticular, and other Trusses, from fixed and moving loads, with practical applications and examples, for the use of Students and Engineers. 87 woodcut illustrations. Fourth edition. 8vo, cloth............$3.50

SHUNK, W. F. The Field Engineer. A Handy Book of Practice in the Survey, Location, and Truck-work of Railroads, containing a large collection of Rules and Tables, original and selected, applicable to both the Standard and Narrow Gauge, and prepared with special reference to the wants of the young Engineer. Ninth edition. Revised and Enlarged. 12mo, morocco, tucks..................$2.50

SIMMS, F. W. A Treatise on the Principles and Practice of Levelling. Showing its application to purposes of Railway Engineering, and the Construction of Roads, etc. Revised and corrected, with the addition of Mr. Laws' Practical Examples for setting out Railway Curves. Illustrated. 8vo, cloth...$2.50

SMITH, ISAAC W., C.E. The Theory of Deflections and of Latitudes and Departures. With special applications to Curvilinear Surveys, for Alignments of Railway Tracks. Illustrated. 16mo, morocco, tucks....$3.00

STILES, AMOS. Tables for Field Engineers. Designed for use in the field. Tables containing all the functions of a one degree curve, from which a corresponding one can be found for any required degree. Also, Tables of Natural Sines and Tangents. 12mo, morocco, tucks $2.00

STONEY, B. D. The Theory of Stresses in Girders and Similar Structures. With observations on the application of Theory to Practice, and Tables of Strength, and other properties of Materials. New revised edition, with numerous additions on Graphic Statics, Pillars, Steel, Wind Pressure, Oscillating Stresses, Working Loads, Riveting, Strength and Tests of Materials. 8vo, 777 pages, 143 illustrations, and 5 folding plates..................$12.50

No. 60.—STRENGTH OF WROUGHT-IRON BRIDGE MEMBERS. By S. W. Robinson, C.E.

No. 61.—POTABLE WATER AND THE DIFFERENT METHODS OF DETECTING IMPURITIES. By Charles W. Folkhard.

No. 62.—THE THEORY OF THE GAS-ENGINE. By Dougald Clerk. Second edition. With additional matter. Edited by F. E. Idell, M.E.

No. 63.—HOUSE DRAINAGE AND SANITARY PLUMBING. By W. P. Gerhard. Fourth edition. Revised.

No. 64.—ELECTRO-MAGNETS. By Th. du Moncel. 2d revised edition.

No. 65.—POCKET LOGARITHMS TO FOUR PLACES OF DECIMALS.

No. 66.—DYNAMO-ELECTRIC MACHINERY. By S. P. Thompson. With notes by F. L. Pope. Third edition.

No. 67.—HYDRAULIC TABLES BASED ON "KUTTER'S FORMULA." By P. J. Flynn.

No. 68.—STEAM-HEATING. By Robert Briggs. Second edition, revised, with additions by A. R. Wolff.

No. 69.—CHEMICAL PROBLEMS. By Prof. J. C. Foye. Second edition, revised and enlarged.

No. 70.—EXPLOSIVES AND EXPLOSIVE COMPOUNDS. By M. Bertholet.

No. 71.—DYNAMIC ELECTRICITY. By John Hopkinson, J. A. Schoolbred, and R. E. Day.

No. 72.—TOPOGRAPHICAL SURVEYING. By George J. Specht, Prof. A. S. Hardy, John B. McMaster, and H. F. Walling.

No. 73.—SYMBOLIC ALGEBRA; OR, THE ALGEBRA OF ALGEBRAIC NUMBERS. By Prof. W. Cain.

No. 74.—TESTING MACHINES: THEIR HISTORY, CONSTRUCTION, AND USE. By Arthur V. Abbott.

No. 75.—RECENT PROGRESS IN DYNAMO-ELECTRIC MACHINES. Being a Supplement to Dynamo-Electric Machinery. By Prof. Sylvanus P. Thompson.

No. 76.—MODERN REPRODUCTIVE GRAPHIC PROCESSES. By Lieut. James S. Pettit, U.S.A.

No. 77.—STADIA SURVEYING. The Theory of Stadia Measurements. By Arthur Winslow.

No. 78.—THE STEAM-ENGINE INDICATOR, AND ITS USE. By W. B. Le Van.

No. 79.—THE FIGURE OF THE EARTH. By Frank C. Roberts, C.E.

No. 80.—HEALTHY FOUNDATIONS FOR HOUSES. By Glenn Brown.